Power Systems

More information about this series at http://www.springer.com/series/4622

Tobias Erlbacher

Lateral Power Transistors in Integrated Circuits

Springer

Tobias Erlbacher
Lehrstuhl für Elektronische Bauelemente
Universität Erlangen-Nürnberg
Erlangen
Germany

ISSN 1612-1287 ISSN 1860-4676 (electronic)
ISBN 978-3-319-00499-0 ISBN 978-3-319-00500-3 (eBook)
DOI 10.1007/978-3-319-00500-3

Library of Congress Control Number: 2014950673

Springer Cham Heidelberg New York Dordrecht London

Printed on acid-free paper

Springer is part of Springer Science+Business Media (www.springer.com)

To my wife, Petra

Preface

This textbook is intended for students and engineers who are engaged in the development of power electronic equipment as well as power amplifiers for radio frequency applications using integrated circuits. It covers relevant technological aspects along the value chain from device physics to power electronic and radio frequency applications.

Reading the corresponding chapters, students will get in touch with device, circuit, and application-specific knowledge to get started on projects in power electronics and radio frequency circuits. The citations throughout this textbook will guide the reader to further literature which provides in-depth description of novel technologies, future applications, and relevant device physics. This textbook thus represents a quick way of getting to the core of device-related topics by following its trails to the technologically relevant literature.

For application engineers, the aim of this textbook is to foster the understanding of device physics and gain insights into new developments of integrated power devices. This understanding is fundamental to properly comparing and choosing fabrication technologies suitable for the respective task at hand, decreasing non-conformity costs in new projects and products and reducing time-to-market.

Process and device engineers will find a decent introduction concerning the requirements of power electronic applications. Moreover, it addresses the needs of process and device engineers that are developing new power electronic devices by motivation of relevant circuit considerations and by reviewing important reliability issues. Its aim is to assist in feasibility considerations when designing these devices and circuits by providing insight into the state of the art of integrated circuit solutions and in presenting current trends and developments that may result in emerging technologies over the next decade. Among the wide range of emerging technologies, those presented in this textbook have high potential to be implemented in future integrated circuit technologies.

The textbook describes power electronic applications using integrated circuits from an academic point of view: Given that a full-fledged overview of applications using integrated circuits is beyond the scope of this book, application examples are selected to provide a conclusive understanding of fundamental requirements inherent to this market.

In summary, the textbook presents a top-down approach from application examples over typical energy conversion and power amplifier circuits down to the lateral power devices, and an in-depth discussion of their ongoing development is given. By review of the present technology on the one hand and of emerging device concepts on the other hand, this textbook is logically divided into two sections: Chaps. 2 through 5 present the current state of the art of applications, circuits and power transistors employed therein. Chapters 6 through 9 explore ongoing developments for lateral power transistors in integrated circuits. Chapter 10 synthesizes the status quo and the emerging technologies to provide insights into potential future applications and solutions.

The introduction presents a brief rundown on the evolution of power electronics and RF applications from an economic perspective. The prospects of these applications in future innovations are discussed, and further demand for the development of the underlying technologies is identified.

In Chap. 2, an introduction to power conversion and RF amplifier applications is presented. After discussion of the impact of these solutions on energy conversion and information technologies, examples of both power electronic and RF amplifier applications are introduced. In order to provide a wide basis for discussion in the subsequent chapters, applications found in a wide range of systems are discussed. Then, the typical requirements for these types of applications are deduced. These requirements form the foundation for the discussion of emerging technologies in the final chapter of this textbook.

Chapter 3 describes circuits for power conversion which are regularly encountered in the aforementioned power electronic applications that can be realized by incorporation of integrated circuits. Moreover, relevant classes of power amplifiers for RF applications are presented here. Discussion of these circuits provides the understanding of requirements that device engineers are typically faced with, when developing new power electronic devices. By combination of the power electronic systems described in Chap. 2 and the circuits used in these applications, important requirements for power semiconductor devices are finally discussed. In order to decide on a suitable fabrication technology, the developer has to understand the benefits and limitations of power devices in integrated circuits in general. Therefore, the requirements are differentiated by considering several applications.

Chapter 4 on power semiconductor devices starts with a summary of so-called "Figures-of-Merit" that are derived from the requirements for systems and circuits listed in Chaps. 2 and 3. These Figures-of-Merit are routinely used to compare different device technologies regarding their applicability for power electronic applications at hand. Next, a rundown of different device topologies that are available for these kinds of applications is provided: The comparison of vertical power MOSFETs, stand-alone RF MOSFETs, and lateral power MOSFETs is used as a basis for the discussion of advantages and limitations when using lateral power MOSFETs in integrated circuits (so-called "smart-power ICs" and "monolithic microwave integrated circuits").

The history of lateral power transistor development presents the first part of Chap. 5 which deals with modern power device technologies in integrated circuits. In particular, the transition from bipolar junction transistors to MOSFETs is described. In this chapter, the operation principle and design considerations of state-of-the-art lateral power MOSFETs and lateral RF MOSFETs are discussed. Moreover, important device concepts like field and ground plates as well as the "RESURF" principle are explained. Finally, the Figures-of-Merit derived in Chap. 4 are evaluated with respect to the implications arising in these lateral power transistors. Device and circuit designers are provided with a toolset to judge capabilities and limitations of different power transistors. Moreover, this state of the art acts as reference for discussions provided in Chaps. 6 through 10.

In Chap. 6, the progress for incorporation of charge compensation patterns is described. In particular, different device designs employing charge compensation and their respective electrical properties are reviewed. Additionally, integration aspects for implementation in smart-power ICs are considered.

Chapter 7 introduced the implementation of trench gates into lateral power transistors. It considers processing technology for trench gate formation, novel device designs, and electrical properties of lateral trench gate power and RF transistors. Again, feasibility of this approach with respect to implementation in integrated circuits is discussed.

The combination of planar and trench gate topologies is considered as an example of "More-than-Moore" integration in Chap. 8 of this textbook. Integration considerations leading to this development are presented and electrical properties of these devices are discussed. The application of this concept for high output power and high frequency operation are reviewed in particular.

Device concepts based on two wide-bandgap semiconductors are described in the Chap. 9 of this textbook. Firstly, the progress on the development of lateral power transistors in silicon carbide is reviewed with respect to power electronics and RF applications. The current state of the art toward the realization of logic circuitry is discussed. Secondly, the integration of gallium nitride high electron mobility transistors on silicon substrates for radio frequency operation represents a promising yet immature development. To judge its potential for application in integrated circuits, the performance of GaN transistors and the requirements for epitaxial layers are reviewed. Moreover, the status on device reliability of GaN transistors is presented.

The final chapter summarizes the aforementioned device concepts considering Figures-of-Merit and integration density. Additionally, process complexity and costs are investigated. To conclude, these developments are rated regarding the application-specific requirements defined in Chap. 2.

This textbook would not have been compiled without the support of several professionals. The author acknowledges the support from Heinz Mitlehner who committed himself to numerous discussions and also arranged for the review of this textbook. Fabrication and analysis of trench gate devices was assisted by Gudrun

Rattmann as well as the technicians at Fraunhofer IISB and the Chair of Electron Devices at the University of Erlangen. Results of scanning spreading resistance measurements of these devices were provided by Florian Stumpf. Moreover, the excellent support during the publication processes by Springer is greatly appreciated.

Erlangen, June

Tobias Erlbacher

Contents

Abbreviations

2DEG	2-Dimensional Electron Gas
AC	Alternating Current
ADSL	Asymmetric Digital Subscriber Line
AGP	Advanced Graphics Port
AlGaN	Aluminum-doped Gallium Nitride
AlN	Aluminum Nitride
ASIC	Application-Specific Integrated Circuit
BCD	Bipolar-CMOS-DMOS (technology)
BiCMOS	Bipolar-CMOS (technology)
BJT	Bipolar Junction Transistor
CAN	Controller Area Network
CAPEX	Capital Expenditures
CMOS	Complementary Metal-Oxide-Semiconductor (technology)
CPU	Central Processing Unit
CTGI	Continuous Trench Gate Integrated (Transistor)
DC	Direct Current
DMOS	Double-diffused Metal-Oxide-Semiconductor
DTI	Deep Trench Isolation
DVB-T	Digital Video Broadcasting-Terrestrial
EDMOS	Extended-Drain Metal-Oxide-Semiconductor
EVDO	Evolution-Data Optimized
FinFET	Finned Field Effect Transistor
FOM	Figure-of-Merit
GaAs	Gallium Arsenide
GaN	Gallium Nitride
GPRS	General Packet Radio Service
GPU	Graphics Processing Unit
GSM	Global System for Mobile Communication
GTO	Gate Turn-Off Thyristor
HBT	Heterojunction Bipolar Transistor

HCI	Hot Carrier Injection
HDD	Hard Disk Drive
HEMT	High Electron Mobility Transistor
HSCSD	High Speed Circuit Switched Data
HSDPA	High Speed Downlink Packet Access
HV-CMOS	High Voltage-Complementary Metal-Oxide-Semiconductor
IC	Integrated Circuit
IGBT	Insulated Gate Bipolar Transistor
ISDN	Integrated Services Digital Network
ITGI	Integrated Trench Gate Integrated (Transistor)
JFET	Junction Field Effect Transistor
LAN	Local Area Network
LDMOS	Lateral Double-diffused Metal-Oxide-Semiconductor
LED	Light Emitting Diode
LIN	Local Interconnect Network
LPCVD	Low Pressure Chemical Vapor Deposition
LTE	Long Term Evolution
MeSFET	Metal-Semiconductor Field Effect Transistor
MMIC	Monolithic Microwave Integrated Circuit
MOSFET	Metal-Oxide-Semiconductor Field Effect Transistor
MTBF	Mean Time Before Failure
nMOS	n-Channel Metal-Oxide-Semiconductor (Transistor)
NPT	Non-Punch Through
OFDM	Orthogonal Frequency-Division Multiplexing
OPEX	Operational Expenditures
PAE	Power Added Efficiency
PAR	Peak-to-Average Ratio
PCIe	Peripheral Component Interconnect Express
pMOS	p-Channel Metal-Oxide-Semiconductor (Transistor)
PT	Punch Through
QV	Quasi-Vertical (LDMOS)
RESURF	Reduced Surface Field
RF	Radio Frequency
S-ATA	Serial Advanced Technology Attachment
SiC	Silicon Carbide
SJ	Superjunction (MOSFET)
SMPS	Switch-Mode Power Supply
SoC	System-on-a-Chip
SOI	Semiconductor on Insulator
SPI	Serial Peripheral Interface
STI	Shallow Trench Isolation
TCAD	Technology Computer Aided Design
TCAS	Traffic Alert and Collision Avoidance System
TGI	Trench Gate Integrated (LDMOS)
TG	Trench Gate (LDMOS)

TTL	Transistor Transistor Logic
UART	Universal Asynchronous Receiver Transmitter
UMOS	Vertical Trench Metal-Oxide-Semiconductor (Transistor)
UMTS	Universal Mobile Telecommunications System
VDMOS	Vertical Double-diffused Metal-Oxide-Semiconductor
VDSL	Very High Speed Digital Subscriber Line
WiMax	Worldwide Interoperability for Microwave Access
WLAN	Wireless Local Area Network

Chapter 1
Introduction

Technological innovation has been the key economic driving force over the last few centuries. Starting with the invention of the steam engine, textiles—mainly made of cotton—became available in large quantities. The production of clothing in factories represents one cornerstone for the industrial revolution in the first part of the nineteenth century. At that time, the invention of the railway already marked the beginning of another economic boost. Heavy investments into the railway system and continuous technology improvement allowed for the installation of a dense track network in the second part of the nineteenth century across Europe and North America. The steam engine driven locomotives enabled mass transportation—both of passengers and goods—long before aircraft or trucks amended the versatility in logistics as we know it today.

Advances in chemistry and electrical engineering such as the introduction of thermodynamic principles in physics and chemistry and discovery of electromagnetic induction and the evolution of the electric engine led to another wave of investments especially towards electricity and electrically powered systems in the first part of the twentieth century. Hydro-electric power plants and steam engines which were the first to produce electricity for local factories were amended by coal-fired power plants and later on by nuclear power plants to cope with increasing demands. The installation of electric grids enabled distribution of electrical power to communities which could then be located further away from power plants.

After the recession marked by Black Tuesday in 1929, another economic boost was evident in the second part of the twentieth century when limitations in mobility of persons and goods were eradicated by the availability of motor vehicles. Here, large investments were also carried out on the road network, crude oil extraction, petroleum production and the installation of filling stations as prerequisites for convenient and everyday usage of cars and trucks.

Following the oil crisis in the 1970s, the economy turned upward with the rise of the semiconductor industry and the mass fabrication of personal computers and mobile telecommunication equipment. Again, significant investments had to be spent for internet and mobile communication infrastructure.

T. Erlbacher, *Lateral Power Transistors in Integrated Circuits*,
Power Systems, DOI 10.1007/978-3-319-00500-3_1

The technological innovations coincided well with the observation of long waves in economic life introduced by Kondratieff in the 1920s for the first three innovation cycles [1, 2]. Based on these observations, Schumpeter has developed a theory on these Kondratieff cycles [3]. Without strictly adhering to the 50 year cycles discovered by Kondratieff, the economic cycles over the last few centuries are schematically illustrated in Fig. 1.1 based on the data from Korotayev [4].

All of these economic cycles have been pushed by severe changes in production methods (manufacturing sites, factories, business complexes) and the methodology of organization (e.g. transportation of goods, persons, energy). Moreover, part of the investments was spent to set up a novel infrastructure (e.g. electric power grids or the internet). This observation is depicted in Fig. 1.2.

It has been identified that investments into infrastructure occur on a shorter time scale of approximately 15–25 years which are called Kuznets swings [4]. Therefore, infrastructure investments occur twice to three times during a typical Kondratieff

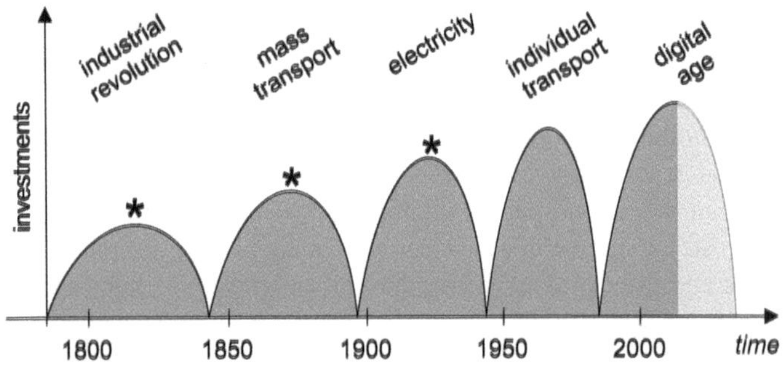

Fig. 1.1 Economic cycles over the last centuries based on the observations from Kondratieff (marked by asterisk) and Korotayev

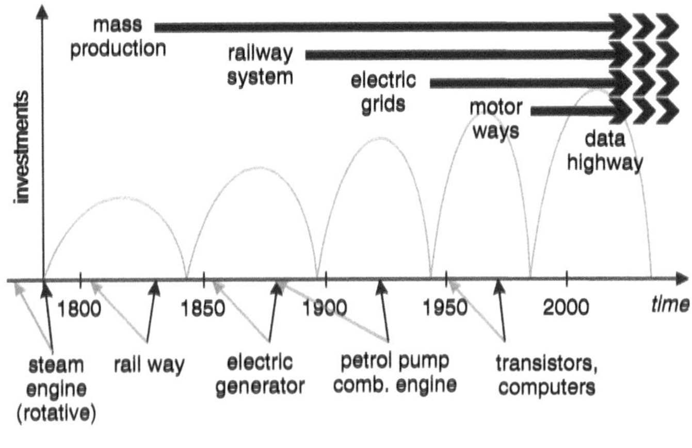

Fig. 1.2 Innovations and infrastructures developed during Kondratieff cycles

cycle. The lifetime of infrastructure, on the other hand, is anticipated to last for 50–70 years. Therefore, development of infrastructure is a decisive factor for implementation of future technologies and the capability of adopting new or improved systems.

Presently, additional investments are being prepared and are starting to be added into the electric grid infrastructure. This includes new DC–DC transmission lines in China and Japan and the strengthening of the existing grid infrastructure in Europe towards the transition to renewable energies. Energy efficient solutions and responsible use of resources have become important subjects in nowadays technological developments. Even though, this effort could be seen as a necessary modernization of existing infrastructure which will not gain enough momentum for another economic boost or change methodologies so significantly on its own.

Today, the digital age appears to be at its verge of investments. With an internet grid allowing for data rates in the Mbit/s range and a mobile phone infrastructure providing the possibility of mobile communication virtually everywhere on this planet, the infrastructure investments into IT appear to be have peaked, which would announce the end of this innovation period.

What kind of innovation will give rise to the next economic cycle—or whether the cycles are significantly speeding up with the broad availability of information technology—is not yet known. But judging from the previous economic cycles it is apparent that groundbreaking innovations were already made and the first demonstrators were fabricated during the previous economic cycle, respectively [4].

From the current innovation perspective, self-diagnostic systems ("Industry 4.0") and health care—especially for the elderly—as well as renewable energies are coming into the focus of governments and companies. The implementation of sensor networks and interconnection of data from different sources are considered to be key prospects for Smart Cities. Whether any of these developments will constitute to another megatrend like the previously described evolutions is not yet clear. According to Rifkin, the next economic cycle, which he calls the "Third Industrial Revolution", will be based on the following five innovations [5]:

- Roll-out of renewable energies
- Buildings as micro power plants
- Energy storage using Hydrogen grids
- Smart power grids
- Electric vehicles for energy storage

With a steady increase in urbanization, investments into new infrastructure for cities are also on the horizon. This includes but is not limited towards the implementation of new food, water and energy distribution systems, monitoring and control of buildings and zero-emission transportation both in mass transit and personal mobility. Assisted and autonomous driving is one way to deal with the increasing traffic density in urban areas. And innovative technologies to reduce traffic are also desired.

The breakthrough of all of these innovations will heavily rely on the availability of energy-efficient systems and a powerful wireless communication infrastructure.

As all of the key innovations boosting the economy so far have also played a vital role in the introduction of subsequent technologies (compare Fig. 1.2), it is safe to assume that power electronic systems will be a key enabling technology for any of these upcoming innovations. Micro- and nanoelectronic circuits are the cornerstone of the present economic cycle. The technological innovations in microelectronics can be subsumed under the compliance to Moore's Law [6]. In order to keep this pace, continuous improvement is required even if Moore's Law cannot be further pursued in the future due to the increasing number of "brick walls" that circuit designers and device technologists are faced with. This includes:

- Electric power conversion systems with high reliability and energy efficiency
- Communication infrastructure with high data rates

As will be discussed in the subsequent chapters, further developments in *Smart-Power ICs* and *RF Power Amplifiers* promise to provide the technological foundation for the ever increasing system demands.

References

1. N.D. Kondratieff, Die langen Wellen der Konjunktur. Archiv für Sozialwissenschaft und Sozialpolitik **56**, 573–609 (1926)
2. N.D. Kondratieff, W.F. Stolper, The Long Waves in Economic Life. Rev. Econ. Stat. **17**(6), 105–115 (1935)
3. J.A. Schumpeter, *The Theory of Economic Development* (Transaction Publishers, New Brunswick, 2004)
4. A.V. Korotayev, S.V. Tsirel, A spectral analysis of world GDP dynamics: Kondratieff waves, Kuznets swings, Juglar and Kitchin Chyles in global economic development, and the 2008–2009 economic crisis. Struct. Dyn. **4**(1), 1–55 (2010)
5. J. Rifkin, *The Third Industrial Revolution: How Lateral Power is Transforming Energy, the Economy, and the World* (Palgrave Macmillan, New York, 2011)
6. G.E. Moore, Cramming more components onto integrated circuits. Electron. Mag. **38**(8), 18 (1965)

Chapter 2
Demand for Power Electronic Systems and Radio-Frequency Applications

This chapter deals with applications for power electronics using integrated circuits, providing intuitive access to lateral power transistors employed therein. For this purpose so-called *smart-power ICs* combining logic operation, memory and power electronic circuitry are presently employed. Additionally, integrated lateral power transistors are also the driving force in multi-stage *RF power amplifiers* (monolithic microwave integrated circuits) for base stations with a clear trend towards further incorporation of logic functionality and the implementation in the power amplifiers of mobile devices.

An overview of power electronic and power amplifier markets provides a basic understanding for the applications driven by these technologies and the implications involved in rolling out new developments. In particular, the impact of power electronics on energy conversion and the impact of power amplifiers on information technologies are presented. Power electronic and RF amplifier applications having a high impact on recent developments using integrated circuits are briefly reviewed. Examples of power electronic applications include motor control and drive inverters for window lifts in automobiles and switch-mode power supplies in battery chargers. Integrated circuits for radio-frequency (RF) applications are readily found in mobile communication base stations and in wireless local area network (WLAN) communication equipment.

Subsequently, the requirements of these applications regarding their commercial use will be summarized. This overview includes economic aspects like energy efficiency, reliability as well as costs of fabrication and operation. Moreover, technologically relevant aspects regarding power density, ruggedness and switching frequency will be elaborated upon.

2.1 Semiconductors in Integrated Power Electronics and Radio Frequency Amplifiers

One of the key enabling technologies for the success of monolithic integrated power electronic systems and RF applications has been the lateral double-diffused metal-oxide-semiconductor (LDMOS) field-effect transistor in silicon technology which

T. Erlbacher, *Lateral Power Transistors in Integrated Circuits*,
Power Systems, DOI 10.1007/978-3-319-00500-3_2

has easily surpassed the performance of integrated bipolar junction transistor in both of these fields.

2.1.1 Impact on Integrated Energy Conversion Systems

Although the first lateral drain extended *Metal-Oxide-Semiconductor Field Effect Transistors* (MOSFET) had already been reported in 1972 [1], and the *reduced surface field* (RESURF) principle was discovered in 1979 [2], it took until the end of the 1990s for commercial integration of LDMOS transistors into complementary MOS (CMOS) technology [3]. Before that, bipolar technology and double-diffused MOS (DMOS) technology using quasi-vertical (DMOS) transistors were the methods of choice to integrate power devices in a logic process [4]. Or from an application point-of-view, power semiconductor devices were first augmented by logic functionality on chip in the 1980s. This augmentation gave rise to the terminology *smart-power integrated circuit* (smart-power IC). At the end of the 1980s, fabrication of display drivers in Bipolar-CMOS-DMOS (BCD) processes began. These monolithic integrated drivers were able to directly drive displays at low current ratings and medium voltages [5]. This market had enough volume to further drive the development of more versatile smart-power IC technology. From the 1990s on, further integrated solutions arose for automotive applications, motor control, power supplies and factory automation equipment. Since that time, non-volatile memories and LDMOS transistors were also employed to improve versatility and ruggedness of power electronic equipment. Advanced options like the implementation of microprocessors, digital signal processors and control functions (e.g. fuzzy logic) became available [4]. The main driver for this development in power electronics, however, was cost-efficiency over similar discrete solutions. Again, large volume production staged the development of these integrated circuits. The timeframe for integrating different power devices into CMOS processes and a comparison to the development of stand-alone Power semiconductor devices is shown in Fig. 2.1.

Besides the bipolar transistor, which has a long track history in BiCMOS process technology, no advanced bipolar power device technologies have been as successful as MOS-based power devices in commercially available smart-power IC processes.

Today, smart-power ICs are part of most electronic equipment with the LDMOS transistors being the working horse among all available power electronic devices. High voltage CMOS (HV-CMOS) technology is particularly interesting due to the lower process complexity compared to BCD technologies. This also simplifies scaling to future technology nodes or the integration of advanced CMOS-related devices like embedded non-volatile memories [7]. Smart-Power ICs found their way into applications where the use of these LDMOS transistors ranges from integrated gate drivers for high power electronic applications using insulated gate bipolar transistors (IGBTs) or superjunction power MOSFETs over battery

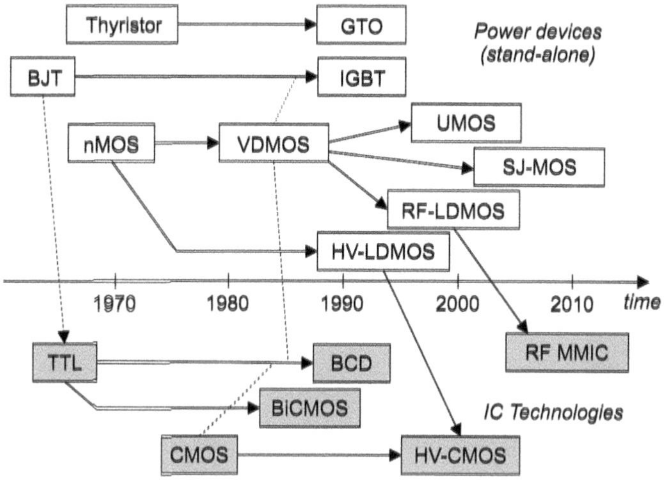

Fig. 2.1 Timeline for development of stand-alone power semiconductor devices and those integrated into CMOS processes, adopted from [4, 6]

monitoring and charging solutions to full-fledged integrated power supplies for on-board or on-chip power conversion. An overview of main-stream applications using smart-power ICs and their respective voltage ratings is given in Fig. 2.2.

Moreover, smart-power ICs are available both as "off-the-shelf", ready-to-use chips and as *application-specific integrated circuits* (ASIC) fabricated on customer demand and even on customer design. The possibility to design individual circuits has given rise to the development of further HV-CMOS, BCD and BiCMOS technologies which silicon foundries are offering to their customers. Now, even for small production volumes these complex manufacturing technologies can be economically employed by sharing a common fabrication process among different customers. Additionally, the ability to reuse design blocks in subsequent

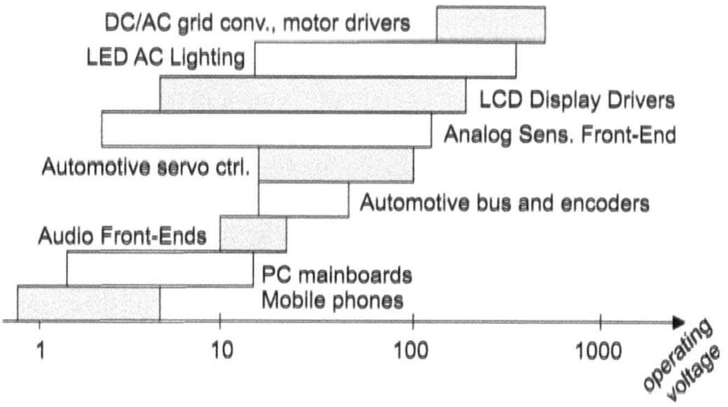

Fig. 2.2 Overview of typical applications employing smart-power IC technologies

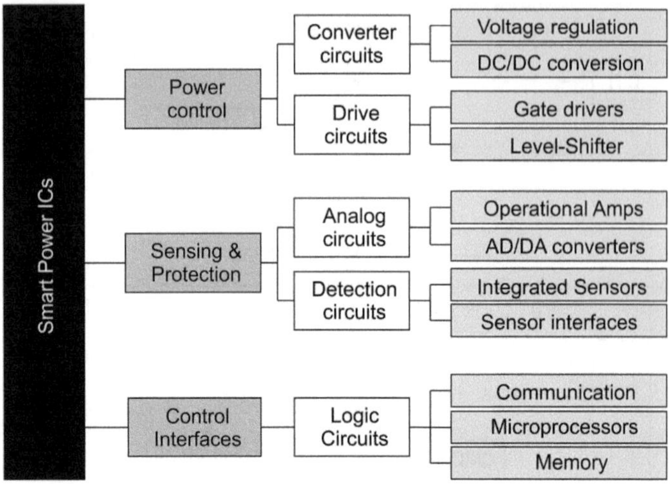

Fig. 2.3 Building blocks regularly encountered in smart-power ICs and RF integrated circuits, adopted from [5]

generations has led to the development of features far beyond simple control circuits for power devices. Typical design features that are included in smart-power ICs can be gathered from Fig. 2.3.

Sensing functionality in smart-power ICs has evolved from over-temperature and over-voltage sensing by on-chip silicon sensors to the integration of detection circuits using external sensors (e.g. humidity, chemical sensors, and optical sensors). The incorporation of high-performance microprocessors, dynamic random access memory and non-volatile memories as well as precision analog-to-digital and digital-to-analog converters has enabled the implementation of small highly specialized computers into a single silicon chip. Most commercially used data transfer protocols (e.g. SPI, UART, I^2C, CAN, LIN, FlexRay) can also be integrated and allow for interaction with the outside world [7]. Often, the semiconductor foundries already offer a base set of these fundamental building blocks to their customers for ease of implementation in new technologies.

Advances from the development of new smart-power IC generations include a reduction in feature size and typically an improvement in device performance which arises from more metallization levels or more robust designs. With each new technology generation becoming available, more sophisticated and powerful solutions are implemented, driving cutting-edge solutions today and in the future. The technologies are based on BiCMOS, BCD or HV-CMOS technologies. A summary of foundry technologies and their typical feature sizes is provided in Fig. 2.4.

With a high investment in the design and development of these integrated circuits, smart-power ICs are often fabricated using a given technology for several years. Some are even fabricated for decades using the same technology node. For instance, some ASICs designed in 2 μm technology are still being fabricated today.

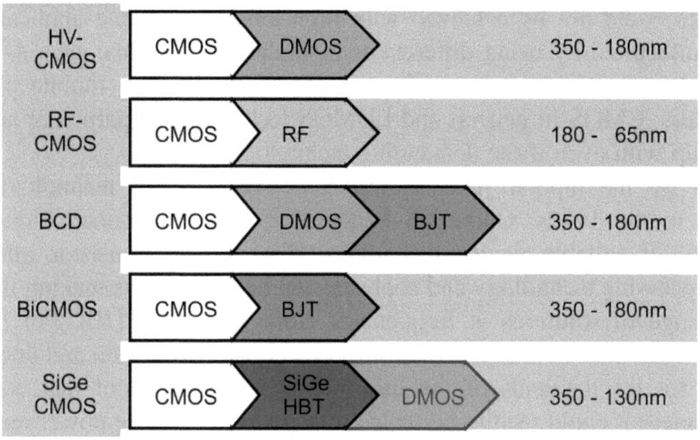

Fig. 2.4 Summary of CMOS-based process technologies available as foundry services and their typical technology nodes

2.1.2 Impact on Information Technologies

Availability of electronic systems capable of reliably operating at frequencies near 1 GHz has enabled the realization of wireless communication infrastructure and radio-frequency applications. A very diverse set of solutions which are part of present day life have evolved from this development ranging from mobile phones and wireless local-area-networks (LAN) over radar applications for avionics to digital video broadcast. An overview of different applications and their operating frequencies is given in Fig. 2.5 along with the year of introduction for the respective standards.

It is apparent that CMOS technology was not available for some of the technologies (e.g. W-LAN) at the year of introduction as the required switching

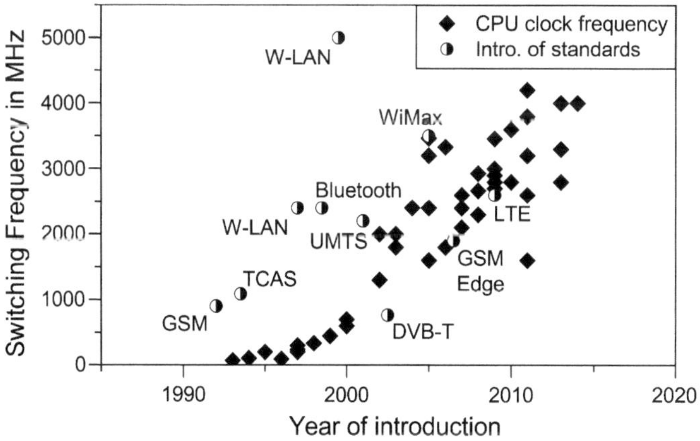

Fig. 2.5 Definition of standards for radio-frequency applications and their operating frequencies by year with comparison to CPU clock frequency by year of introduction to market

frequencies could not be achieved with high efficiency. Some applications have been manufactured by using different semiconductor materials, such as the GaAs heterojunction bipolar transistors for the power amplifier in mobile phones [8]. Nevertheless, CMOS in general and LDMOS technology in particular are recently catching up with even these demanding technologies [9].

Driven by the bipolar junction transistor (BJT), these technologies started emerging in the 1970s. Operating frequencies were, of course, limited by the availability of suitable device technology. The bipolar transistor offered quite mature processing technology and sophisticated design options making it viable for a wide range of solutions at frequencies close to 1 GHz [10]. For the simple modulation schemes of first generation RF power amplifiers, gain and linearity were sufficient for the demands of the information technologies of that era, and the bipolar transistor could readily be scaled to support the diverse power requirements of different applications [11].

Starting in the 1990s, the technology for LDMOS transistors tailored towards high operating frequencies had evolved so efficiently, that it started to replace the silicon bipolar transistor in radio-frequency applications [12]. A comparison between bipolar junction transistors and unipolar MOS transistors can be found in Chap. 4 of this textbook. Furthermore, higher operating frequencies and more sophisticated modulation schemes have been realized due to the high gain at better linearity of these unipolar RF transistors [13, 14]. This disruptive technology paved the way for today's third and fourth generation RF power amplifiers in mobile communication base stations. It has also enabled a tremendous leap in data transmission rates using wireless telecommunication equipment. A comparison to data transfer rates (downstream) with the year of introduction to the European market is given in Fig. 2.6.

As is evident from the data, mobile phones are now capable of providing similar data transfer rates to cable or DSL-based internet access. Presently, RF-LDMOS

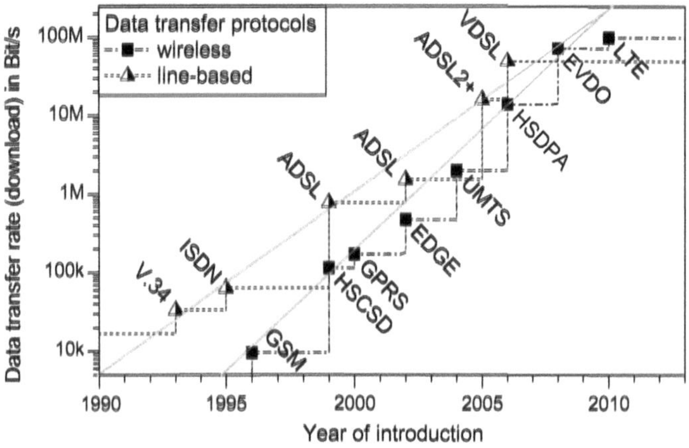

Fig. 2.6 Data transfer rates (downstream) of wireless and line-based telecommunication technologies and year of introduction into the European market

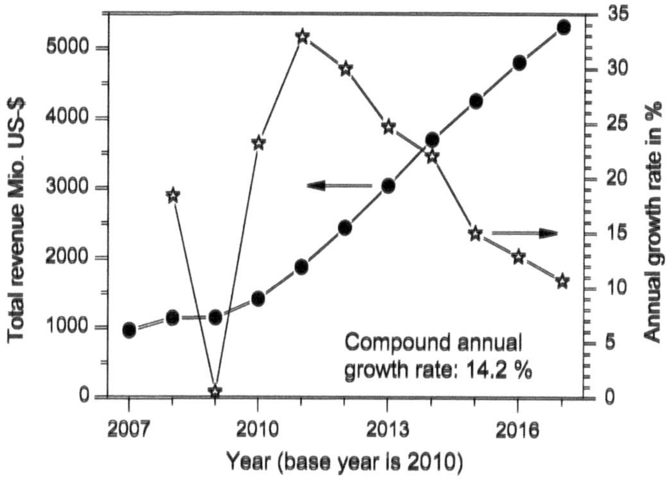

Fig. 2.7 Estimated market development for mobile communication RF amplifiers, adopted from [15]

transistors are the key element of modern mobile phone base stations. Here, market prospects are excellent, and a compound annual growth rate of up to 30 % is predicted for the next several years as depicted in Fig. 2.7.

The market volume for these RF amplifiers is expected to arrive at over 5 billion US-$ in 2017. This includes a share of 75 million US-$ for RF semiconductor devices [15]. The key drivers for RF semiconductor devices are the demand for new features, the increasing deployment of 3G and 4G mobile phones, and the penetration of new markets (e.g. LTE). Figure 2.8 summarizes the technologies currently used in RF amplifiers depending on power level and operation frequency.

Just recently, competitive performance of integrated silicon LDMOS transistors for application on mobile phone power amplifiers was reported, a terrain that has been traditionally occupied by GaAs transistors [17]. Similarly, the efficient operation of the first CMOS-based power amplifiers integrated into silicon for WiMax and LTE transmitters has been achieved [18]. For mobile phones, monolithic integration of circuits in system-on-chip solutions today is state-of-the-art. Manufacturers strive for a single-chip solution using a CMOS-based technology for cost reasons [15, 19]. The main advantages of using power devices for the RF power amplifier with operating voltages well above the CMOS level (typically 3.3 V or below) arises from the reduction of device current for the required output power of 1–3 W. This would both minimize electromigration effects through the metallization layers and the impact of parasitic inductances [20]. Therefore, it can be expected that the trend towards a fully integrated fabrication with LDMOS transistors will also be used in RF power amplifiers for mobile phones in the near future. This is particularly true, because the CMOS technology alone is not capable of providing cost-efficient RF power amplifiers for handsets due to diminishing returns from technology scaling [21].

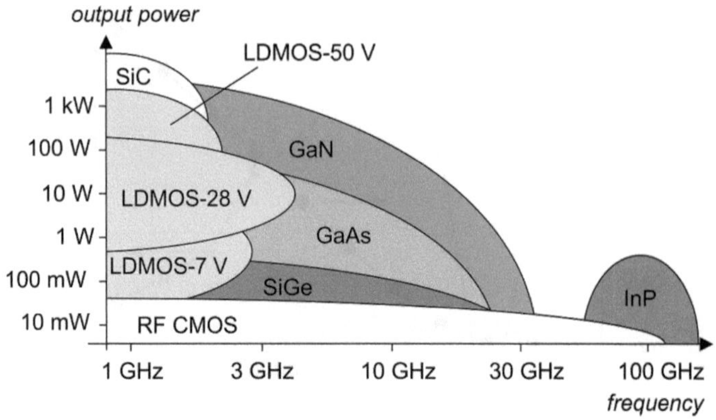

Fig. 2.8 Dominant semiconductor technologies currently used in RF power amplifiers (adopted from [16])

The implementation of monolithic microwave integrated circuits (MMICs) allows the concatenation of several amplifier stages on a single substrate. This minimizes parasitic capacitances and inductances and can offer higher output power.

2.2 Integrated Power Electronic and Radio-Frequency Applications

Monolithic integration of power semiconductor devices and the implementation of advanced circuit topologies have enabled a wide range of power electronic applications. Selected examples demonstrate the benefits of this approach.

2.2.1 Switch-Mode Power Supplies for DC-to-DC Conversion

A high efficiency well beyond 90 %, high tolerances regarding input voltage and grid frequency and a low weight and volume are the key benefits of switch-mode power supplies (SMPS) compared to fixed voltage regulators. Figure 2.9 depicts the building blocks of an SMPS.

For high power densities like in traction systems vertical power transistors (MOSFETs and IGBTs) on transfer substrates with water cooling for efficient heat management are usually employed. Low and medium power densities use packaged power devices and passive (air) cooling that can be augmented by a heat sink. The monolithic integration using lateral transistors together with the high energy efficiency of this circuit topology provides the basis for control, routing and regulation of medium power densities in smart-power ICs.

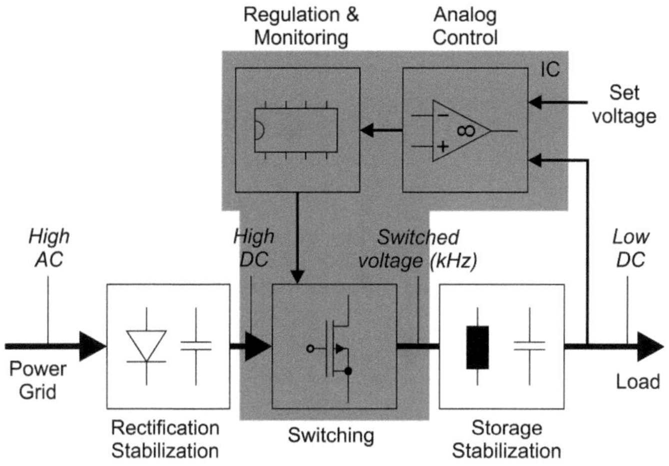

Fig. 2.9 Circuit topology of monolithic integrated switch-mode power supply

2.2.2 Motor Control and Drive Inverters for DC-to-AC Conversion

Traditionally, most three-phase induction motors are operated directly on the AC line at a fixed speed related to the net line cycle. For applications with varying loads, e.g. water pumps or ventilation in air conditioning, a higher energy-efficiency can be achieved by variable frequency drives. Using power electronics, the torque and speed of the electric motor can be directly controlled by supplying optimized voltages and currents to the motor. This is performed by converting the input power of this drive inverter to a sinusoidal wave with power adequate for the present demand of the motor. A schematic circuit topology for a DC-to-AC converter is shown in Fig. 2.10.

This topology can generally be used to convert direct current to alternating current. A "mobile AC plug" powered by the battery of an electric or hybrid vehicle is an example for this type of energy conversion.

Due to its high reliability, low fabrication effort (at high volumes) and high energy efficiency, monolithically integrated drivers for bidirectional variable frequency drives are applied in automotive applications, especially for power window lifts. A block diagram of such a power window lift using an integrated circuit is shown in Fig. 2.11.

2.2.3 Power Amplifiers for Mobile Communication Base Stations and Handhelds

One of the largest growing markets of the last and present decades is telecommunications enabling access to the internet. The introduction of higher data rates

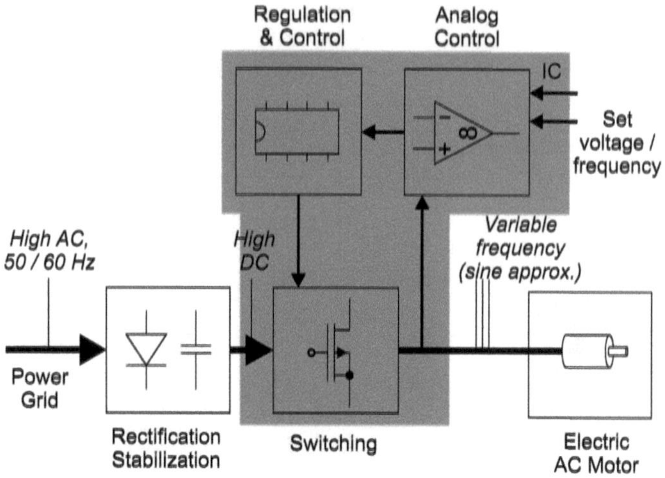

Fig. 2.10 Circuit topology of monolithic integrated variable frequency drive

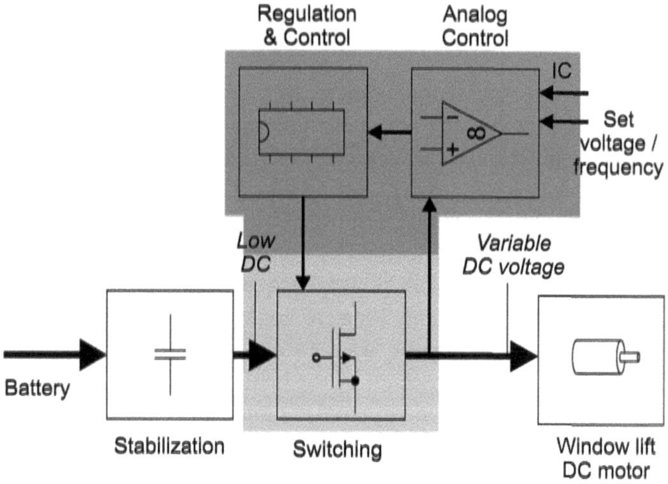

Fig. 2.11 Circuit topology of monolithic integrated power window lifts circuitry

has led to computer technology "to go" with laptops and smart phones on the forefront. At the core of this development is the transmission of information via radio waves in the lower GHz range. The underlying transmission infrastructure is schematically depicted in Fig. 2.12.

It consists of both a sender (signal source, e.g. microphone) and receiver (signal sink, e.g. loudspeaker). The signal source is encoded in the baseband processing hardware that typically consists of a digital signal processor and an ASIC [22]. Then, a radio frequency frontend is used to modulate the digital signal onto the

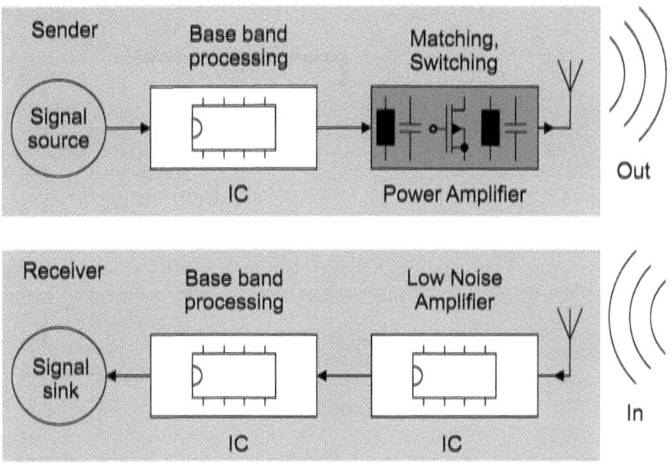

Fig. 2.12 Schematic representation of transmission link in telecommunications

carrier frequency. This modulated carrier signal is amplified by the RF power amplifier and broadcast using an antenna.

An RF power amplifier in base station equipment utilizes silicon RF LDMOS transistors and offers a peak output power of 120–180 W at a supply voltage of 26–28 V and frequencies of 900 MHz (GSM) to 3.6 GHz (UMTS) depending on the baseband technology. The power amplifier in the mobile phone is still fabricated using GaAs HBT transistors in MMIC technology [17].

When the transmitted signal is received, it is passed through a low noise amplifier to extract the signal from the hostile background noise: Assuming that the power of the received signal compares to the size of a pin (approximately one mm in diameter), the size of the haystack—given by the largest known interferer—would be the size of a field of 100 by 100 m [23]. Fortunately, the interferences occur at other frequencies than the desired signal and can be filtered out. The amplified signal can then be passed on to be decoded in the baseband processing hardware, where synchronization and data recovery take place [24]. This functionality can be clustered into a single silicon CMOS chip.

2.2.4 Transceivers for Wireless-LAN Communication

For the connection of mobile phones, laptops, desktop computers and peripherals through a local area network, wireless LAN was implemented following the IEEE 802.11 protocol [25]. The purpose of this protocol is to provide a communication frameset that achieves data rates similar to wire bound LAN connections for up to 52 potential clients. Usually, all clients are connected to a wireless access point. The wireless access point is often integrated into a router providing connection to a

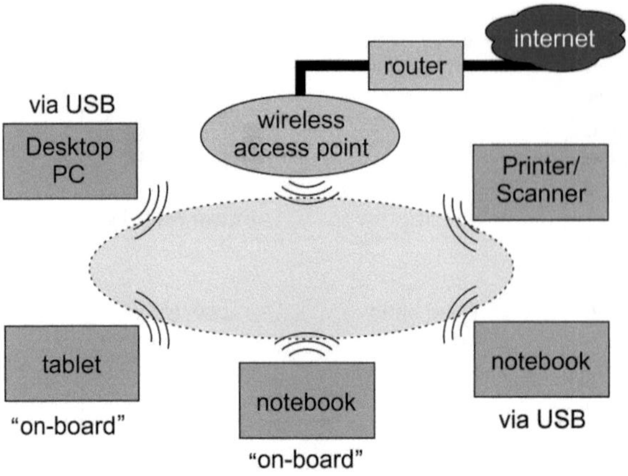

Fig. 2.13 System architecture for a wireless LAN with one central access point and participants connected through different hardware components

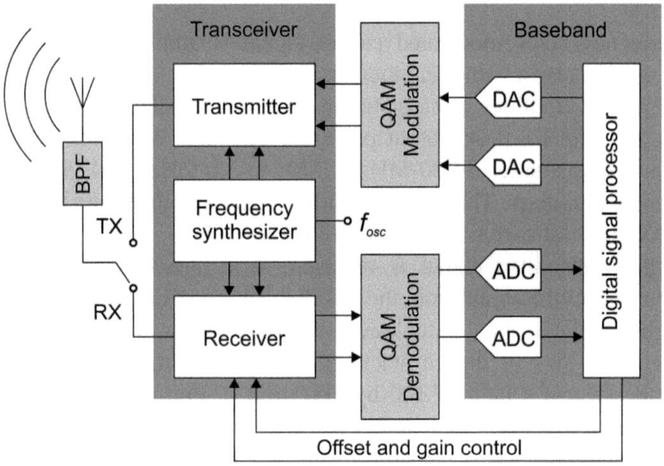

Fig. 2.14 Block diagram of a wireless LAN transceiver operating at 5 GHz (adopted from [26])

wire-bound intranet or the internet (e.g. DSL or cable TV). Figure 2.13 represents a system architecture regularly encountered in households or small companies.

Moreover, mobile phones with integrated wireless LAN functionality provide internet connectivity using wireless telecommunication data transfer (e.g. UMTS, LTE) in ad hoc networks. The block diagram for a wireless LAN transceiver is shown in Fig. 2.14.

In order to ensure that the information from up to 52 potential clients—represented by 52 subcarriers of the orthogonal frequency-division multiplexing (OFDM) signal—can be transmitted without distortion, the transmitter stage has to provide a peak-to-average ratio of at least 17 dB. This requires high power back-off in the transmitter and a wide dynamic range in the receiver [26]. Owing to error correction topologies, these demanding constraints can be somewhat relaxed.

Due to the limited output power (of up to 1 W depending on the country) that is allowed for a wireless LAN, CMOS technology can be applied for the implementation of RF power amplifiers in wireless LAN transmitters, and integrated circuits are routinely used. Nevertheless, to boost the signal level of the clients to the access point, external power amplifiers can be hooked up to the CMOS output stage [27].

2.3 Requirements for Power Electronic and RF Amplifier Systems

2.3.1 *Energy Efficiency*

In modern power electronic systems, energy efficiency can be a key criterion for customers when buying products. This is particularly true for mobile systems where "infinite" power supply from a wall plug is not available. Also, this point holds for applications like white goods where a rating system with respect to energy efficiency has been implemented. The customer is made aware of the system's efficiency, and the system facilitates low operating expenditures. Or, in the case of solar inverters, the revenue is increased from using more efficient equipment despite the higher initial costs [28].

Moreover, efficient energy usage produces less heat. Subsequently, the cooling system can be reduced in size and cost. This advantage becomes particularly apparent, when air instead of water cooling can be applied.

2.3.2 *Size and Weight of Equipment*

Reducing the size and weight of a power electronic system is particularly beneficial in mobile applications where large, bulky or heavy equipment is not accepted by the customer, e.g. mobile phones. While power semiconductor devices are usually not the constraint in size and weight of electronic equipment, they may affect these properties of related subassemblies. Two prominent examples are passives components and cooling systems.

- By applying higher switching frequencies in power converters, smaller inductances and capacitances are required.
- The usage of more efficient semiconductor devices allows for a reduction in the size of the cooling system. With a large benefit arising, when air cooling instead of a fluid cooling system is sufficient to remove the excess heat.

2.3.3 Power Density

From an application perspective, the power density of an energy conversion system depends on a wide range of factors. Passive components like inductors and capacitors typically require a lot of volume. Moreover, cooling components (heat sinks, pipes, fluids etc.) constitute to another mentionable portion. In contrast, the (power) semiconductor devices are not a key factor when power density of the system is concerned. However, as shown in Chap. 4, the power density of semiconductor devices is of concern, when chip costs have to be considered.

2.3.4 Reliability

Another important aspect of electronic equipment is reliability. It is defined as "the ability of a system to consistently perform its intended or required function or mission, on demand and without degradation or failure" [29]. In Industrial, Science and Medical applications, failed systems can usually be repaired. In this case, the reliability of a system can be estimated by the "mean-time-between-failures" (MTBF). When an electronic subsystem has failed, it is typically replaced by a new one, unless repair of the electronic component requires less labour and parts than the full replacement. Additionally, failure of power electronic equipment often causes irreparable damage to the whole subsystem. Then, the reliability of a system is usually predicted by the "mean-time-to-failure" (MTTF) which is equivalent to the total lifetime. The reliability requirements greatly vary between different markets, e.g. lighting on the lower end and aviation on the upper end of reliability demands. The failure probability of electronic equipment can be described using a "bathtub curve" (see Fig. 2.15).

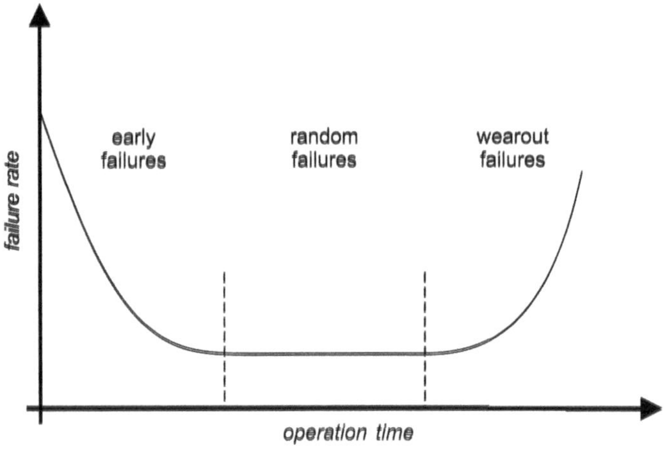

Fig. 2.15 Bathtub curve describing failure probability of electronic equipment

This curve predicts the end of a product's life during one of three different periods [30]:

- A high initial failure probability with a subsequent decline of failure rate. The term "infant mortality" refers to the first observation of this bathtub-like behaviour in human life expectancy [31].
- A low constant failure probability during the "useful life" of the product.
- An increase in failure probability occurs due to wear-out and its onset typically defines the usual operation time of electronic equipment.

It must be noted that this behaviour is not observed for all electronic systems, and that the applicability on a system level is at least questionable [32]. However, in the context of this textbook which focuses on power semiconductor devices and monolithic integration, the bathtub curve for modelling of device failures is well accepted. Early failures are routinely identified by burn-in tests. By measuring and modelling the wear-out region, it is possible to design electronic equipment for a specified operating lifetime with failure rates below 1 ppm (parts-per-billion). Moreover, doing so allows for the prediction of expected failures in the field. This is particularly important for markets where excessive failures cannot be tolerated (spaceflight) or preventive repair must be performed due to safety regulations (automotive). Of course, other markets are not as demanding of highly reliable equipment, e.g. consumer applications.

In microelectronics, device failure follows a Weibull failure distribution [33], which is a monotonic function. In order to apply the Weibull theory to the non-monotonic bathtub function, different modifications of the classical Weibull equation have been proposed [34].

2.3.5 Time-to-Market

In fast evolving markets like the mobile communications sector, the ability to be the "first to deliver" can make the difference between making a profit or making a loss. Therefore, time-to-market is an important factor when implementing new designs. Generally, development cycles for integrated circuits are more time consuming when a new design needs to be fabricated and tested. This is particularly true for RF amplifiers, where redesigns are frequently necessary to keep up the pace [21]. Under these conditions, where fully Integrated System-on-Chip solutions become rapidly outdated, it can be favourable to have a solution with different sub-circuits where key components can be more easily changed. In contrast, wireless LAN applications or data transmission subsystems (e.g. using Bluetooth) are well established and do not readily change. Generally spoken, the benefit of a fast production ramp-up with previous engineering runs, the capability of using parts of previous designs and the availability of excellent modelling tools for all kinds of applications have turned the implementation of integrated circuits into a fast time-to-market approach.

2.3.6 System Cost

From an economic perspective, cost is a key factor when designing new systems. The cost issue is twofold in nature: The first consideration is capital expenditures (CAPEX) which relates to the installation price for the customer. This leads to careful calculation of fabrication, packaging and assembly costs necessary to derive cost-efficient solutions. These calculations then give a first indication on the feasibility of monolithic integration of power electronic systems. Unfortunately, there are other risks that cannot be easily converted into monetary considerations. For example, the implementation of a more reliable technology could be feasible even if it is more expensive. To this respect is should be noted that the costs for additional rework or service in the field are often difficult to estimate for companies along the whole value chain. Moreover, a reliable measure beyond previous experience within the company for customer satisfaction or recalls often does not exist. The second consideration is operational expenditures (OPEX) which contains the costs for operating the equipment. These recurring costs are particularly relevant in systems which consume a lot of energy or that are expensive to repair, e.g. in down-hole drilling [35].

Most conventional power electronic systems which operate in the range of up to 100 W could be integrated using smart-power ICs. However, the feasibility of doing so strongly depends on these economic considerations. Even though smart-power ICs achieve a higher reliability than circuits assembled from discrete devices, it is argued that the partitioning of logic, memory and power devices within a smart-power IC must be suitable to warrant integrated circuit design. If the additional cost for packaging and assembly still favours a lower cost for a discrete system, the monolithic integration of this particular circuit will usually be discarded [5]. Taking this to the extreme, monolithic integration of a drive inverter for an electric car using LDMOS transistors for the power switches is currently not cost efficient. This is, however, not so say that such a drive inverter does not employ smart-power ICs in the form of gate drivers etc.

2.3.7 Switching Frequency

For power electronic systems, an increase in switching frequency allows for smaller passive devices (e.g. inductance of a DC-DC converter). Given that switching frequency is typically limited by the parasitic components of the packaging and assembly techniques and not of the power switches themselves, it can be conceived that a high switching frequency can be achieved by monolithic integration in smart-power ICs. While the operation at RF frequencies is not of relevance when designing energy conversion circuits, there is a clear trend observable towards higher operating frequencies in order to reduce the size and weight of the electronic system.

A significantly stronger strive for increased operating frequencies holds true for radio-frequency applications where overall system performance strongly depends on the choice of a suitable amplifier design. This development is also driven by military applications, where higher operating frequencies enable secure and uninterruptable communication.

2.3.8 Ruggedness

Finally, ruggedness determines the systems capability to withstand unexpected operation conditions. For example, these may arise from supply voltage fluctuations due to lightning strikes or a short-circuited load from a failing electric engine. In order to prevent excessive damage, these faults have to be detected and the electronic equipment must be quickly switched to a self-protection mode. A rugged system which is capable of surviving these kinds of faults requires both fast intelligent control circuitry and power devices that are not destroyed between the occurrence and detection of the fault.

2.3.9 Comparison of Requirements

The correlations between these requirements for power electronic systems and radio-frequency broadcasting equipment and their relevance for different applications are summarized in Table 2.1.

Table 2.1 Relevance of properties for power electronic systems and radio-frequency broadcasting equipment and their relevance depending on selected markets

Power electronic systems				
	Automotive	UPS	Consumer	Industrial
Cost	Very high	High	High	High
Size	High	Low	Low/Medium[a]	Low/Medium[a]
Efficiency	High	High	Medium	High
Frequency	Low	Low	Low	Low/Medium[a]
Time-to-market	Medium	Medium	Very high	High
Reliability/Rugged.	Very high	Very high	Medium	Medium/High[a]
Radio-frequency broadcasting equipment				
	Base station	Mobile phones	Office/Business	Aerospace
Cost	High	High	High	Medium
Size	Medium	High	Medium	Low/High[a]
Efficiency	Medium	High	Medium	Medium
Frequency	Very high	Very high	Very high	Very high
Time-to-market	High	Very High	High	Low
Reliability/Rugged.	High	Medium	Medium	Very high

[a] depending on application

These requirements need to be considered when designing a competitive power electronic system. However, a significant proportion of these requirements depend on the particular market a system is designed for. Similar solutions may have different constraints coming from the customer base across countries (national customs and trends), competition and legislation (laws and regulations).

References

1. H.J. Sigg, G.D. Vendelin, T.P. Cauge, J. Kocsis, D-MOS transistor for microwave applications. IEEE Trans. Electron Devices **9**(1), 45–53 (1972)
2. J.A. Appels, H.M.S. Vaes, High voltage thin layer devices, in *International Electron Devices Meeting* (Washington, USA, 1979)
3. J.L. Sanchez, State of the art and trends in power integration, in *MSM*, Puerto Rico, USA, 1999
4. B. Murari, C. Contiero, R. Gariboldi, S. Sueri, A. Russo, Smart power technologies evolution, in *IEEE Industry Applications Conference* (Rome, Italy, 2000)
5. B.J. Baliga, An overview of smart power technology, IEEE Trans Electron Devices **38**(7), 1568–1575, (1991)
6. G. Majumdar, Future of power semiconductors, in *Power Electronics Specialists Conference* (Aachen, Germany, 2004)
7. M. Schrems, M. Knaipp, H. Enichlmair, V. Vescoli, R. Minixhofer, E. Seebacher, F. Leisenberger, E. Wachmann, G. Schatzberger, H. Gensinger, Scalable high voltage CMOS technology for smart power and sensor applications. Elektrotechnik Informationstechnik **125**(4), 109–117 (2008)
8. T. Ishibashi, Y. Yamauchi, A possible near-ballistic collection in AlGaAs/GaAs HBT with a modified collector structure. IEEE Trans. Electron Devices **35**(4), 401–404 (1988)
9. P.J. Zampardi, Will CMOS amplifiers ever kick-GaAs?, in *IEEE Custom Integrated Circuits Conference* (San Jose, USA, 2010)
10. C.P. Snapp, Microwave bipolar transistor technology—present and prospects, in *European Microwave Conference* (Brighton, United Kingdom, 1979)
11. W. Burger, C. Dragon, in *Silicon LDMOS and VDMOS transistors—Physics, design and technology*, ed. by John L. B. Walker. Handbook of RF and Microwave Power Amplifiers (New York, Cambridge University Press, 2012), pp. 1–39
12. N. Camilleri, J. Costa, D. Lovelace, D. Ngo, Silicon MOSFETs. The microwave device technoloclv for the 90s, in *MTT-S International Microwave Symposium*, pp. 545–548, 1993
13. A. Wood, W. Brakensiek, C. Dragon, W. Burger, 120 Watt, 2 GHz, Si LDMOS RF power transistor for PCS base station applications, in *IEEE MTT-S International Microwave Symposium* (Baltimore, USA, 1998)
14. K. Shenai, E. McShane, Current status and emerging trends in RF power FET technologies, in *MTT-S International Microwave Symposium* (Phoenix, USA, 2001)
15. Frost and Sullivan, World RF Handset Semiconductor Market, N8D6-26, 2011
16. S.J.C.H. Theeuwen, H. Mollee, LDMOS Transistors in power microwave applications, *White paper at Microsoft Journal Web*, p. 5, 2008
17. A. Tombak, D.C. Dening, M.S. Carroll, J. Costa, E. Spears, High-efficiency cellular power amplifiers based on a modified LDMOS process on bulk silicon and silicon-on-insulator substrates with integrated power management circuitry. IEEE Trans. Microwave Theory Tech. **60**(6), 1862–1869 (2012)
18. Y. Li, J. Lopez, P.-H. Wu, W. Hu, R. Wu, D.Y.C. Lie, A SiGe envelope-tracking power amplifier with integrated CMOS envelope modulator for mobile WiMAX/3GPP LTE transmitters. IEEE Trans. Microwave Theory Tech. **59**(10), 2525–2536 (2011)

19. M.S.J. Steyaert, J. Janssens, B. de Muer, M. Borremans, N. Itoh, A 2-V CMOS cellular transceiver front-end. IEEE J. Solid-State Circ **30**(10), 1895–1907 (2000)
20. J. Ramos, M. Steyaert, High voltage devices for RF power amplifiers: an advantage?, in *Analog Circuit Design* (New York, Springer US, 2004), pp. 177–200
21. D.Y.C. Lie, "RF-SoC": integration trends of on-chip CMOS power amplifiers: benefits of external PA versus integrated PA for portable wireless communications. Int. J. Microwave Sci. Technol. **2010**, 380108 (2010)
22. Y. Hawwar, E. Farag, S. Vanakayala, R. Pauls, X. Yang, S. Subramanian, P. Sadhanala, L. Yang, B. Wang, Z. Li, H. Chen, Z. Lu, D. Clark, T. Fasket, P. Mallela, M. Shelton, D. Laurens, T. Salaun, L. Goegeon, N. Aubourg, H. Morvan, N. Le Henaff, G.F. Prat, F. Charles, C. Creach, Y. Creach, C.Y. Calvez, P. Butel, 3G UMTS wireless system physical layer: baseband processing hardware implementation perspective. IEEE Commun. Mag. **44**(9), 52–58 (2006)
23. P.C. Davis, Merits and requirements of a few RF architectures, in *Bipolar/BiCMOS Circuits and Technology Meeting*, pp. 62–66, 26–28 Sep 1999
24. L.E. Larson, Silicon technology tradeoffs for radio-frequency/mixed-signal "Systems-on-a-Chip". IEEE Trans. Electron Devices **50**(3), 683–699 (2003)
25. IEEE Standard for Information technology–Telecommunications and information exchange between systems Local and metropolitan area networks–Specific requirements Part 11: Wireless LAN Medium Access Control (MAC) and Physical Layer (PHY) Specifications, *IEEE Std 802.11-2012*, pp. 1–2793, 29 Mar 2012
26. M. Zargari, D.K. Su, C.P. Yue, S. Rabii, D. Weber, B.J. Kaczynski, S.S. Mehta, K. Singh, S. Mendis, B.A. Wooley, A 5-GHz CMOS transceiver for IEEE 802.11a wireless LAN systems. IEEE J. Solid-State Circuits **37**(12), 1688–1694 (2002)
27. M. Zargari, M. Terrovitis, S.H.-M. Jen, B.J. Kaczynski, M. Lee, M.P. Mack, S.S. Mehta, S. Mendis, K. Onodera, H. Samavati, W.W. Si, K. Singh, A. Tabatabaei, D. Weber, D.K. Su, B.A. Wooley, A single-chip dual-band tri-mode CMOS transceiver for IEEE 802.11a/b/g wireless LAN. IEEE J. Solid-State Circuits **39**(12), 2239–2249 (2004)
28. B. Burger, D. Kranzer, Extreme high efficiency PV-power inverters, in *European Conferene on Power Electronics and Applications* (Barcelona, 2009)
29. International organization for standardization, *Quality management systems—Fundamentals and vocabulary (ISO 9000:2005), Genève* (International, Organization for Standardization, Switzerland, 2005)
30. B. Roesch, *Reliability, in handbook of RF and microwave power amplifiers* (Cambridge University Press, Cambridge, 2012), pp. 446–507
31. E. Halley, An estimate of the degrees of the mortality of mankind, drawn from curious tables of the birth and funerals at the city of Breslau; with an attempt to ascertain the price of annuities upon lives, Phiolophical Trans. Royal Soc. Lond. **17**, 596–610, 1693
32. G.-A. Klutke, P.C. Kiessler, M.A. Wortman, A critical look at the bathtub curve. IEEE Trans. Reliab. **52**(1), 125–129 (2003)
33. W. Weibull, a statistical distribution function of wide applicability. ASME J. Appl. Mech. **18**, 293–297 (1951)
34. M. Bebbington, C.D. Lai, R. Zitikis, A flexible Weibull extension. Reliab. Eng. Syst. Safety **92**(6), 719–726 (2007)
35. S. Zhan, I. Ahmad, L. Heuermann-Kuehn, J. Baumann, B. Hughes, Integrated PoF and CBM strategies for improving electronics reliability performance of downhole MWD and LWD tools, in *SPE Annual Technical Conference and Exhibition* (Florence, Italy, 2010)

Chapter 3
Power Electronic and RF Amplifier Circuits

The applications mentioned in Chap. 2 require suitable power electronic circuits. This chapter transfers the application specific constraints and requirements down to the circuit and device level. Power electronic systems using integrated circuits are enabled by circuitry for energy conversion and control. Here, switch mode converters provide high efficiency, simple topology and high power densities satisfying the requirements set by the application. The benefits and drawbacks of switch-mode conversion in comparison to conventional fixed voltage regulation are explained. Moreover, the topologies for buck and boost inverters in direct current conversion and solid-state rectification as well as for power inverters using half-bridges are reviewed with respect to the application demands.

Next, fundamentals of circuits commonly found in state-of-the-art RF amplifiers are also introduced. The benefits and limitations of different amplifiers types (classes) are discussed in order to foster understanding of those suitable for the application specific requirements.

Using these circuit topologies, the application-specific requirements for power transistors in power electronic applications are discussed. For different circuits, requirements regarding energy efficiency, power density and switching frequency are aspects that each power semiconductor device has to satisfy to a different degree.

3.1 Circuits for Energy Conversion and Control

Powering a load in an electric circuit requires that a suitable form of electricity is provided. A light emitting diode or a light bulb demands a different source of electricity than an asynchronous motor. The electric energy can be provided at different *frequencies* ranging from constant direct current to high frequencies of alternating current for motor control. Moreover, the *voltage* and *current* consumed by the load varies over a wide range from nanowatts for the power supply of integrated circuits and sensors over kilowatts when driving electric engines to gigawatts in femtosecond pulsed lasers [1].

T. Erlbacher, *Lateral Power Transistors in Integrated Circuits*,
Power Systems, DOI 10.1007/978-3-319-00500-3_3

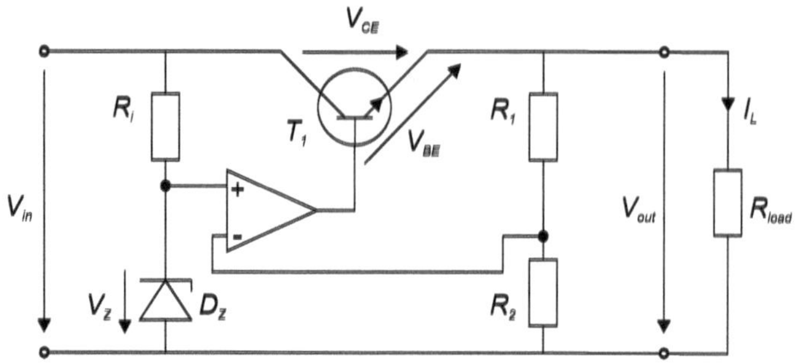

Fig. 3.1 Fundamental circuit topology of fixed voltage regulator, adopted from [2]

Another point to consider is the direction of energy flow. While most loads like lighting, heating and cooling systems simply consume power (*unidirectional*), there are applications that require a *bidirectional* flow of current. Examples are the control circuitry for charging and discharging secondary batteries and the motor of an electric vehicle when recuperation is desired. In order to transform the energy for a given load, power electronic circuits have been developed.

A very simple yet inefficient method of converting the input voltage of an electronic circuit to the required voltage level of a load at direct current is the application of a fixed voltage regulator as shown in Fig. 3.1.

Today, fixed voltage regulators are routinely employed in electronic circuits. However, these circuits exhibit several drawbacks like a low power factor and poor energy-efficiency [3].

3.1.1 Switch-Mode Converter Circuits

An advanced methodology incurring lower conversion losses that has successfully emerged in energy conversion is the implementation of switch-mode topologies. The main concept of any switch mode circuitry is the transportation of energy in small portions from the primary to the secondary side at sufficiently high switching frequencies. In bidirectional topologies, energy flow from the secondary to the primary side is also possible. Typically, an inductor is used to store the energy portion, whereas a capacitor is used to store the energy at the primary and secondary side, respectively [4]. The first pulse-width controlled AC-to-DC converter for single and three phase AC currents was described in 1970 [5]. From this idea a large number of different converter topologies have evolved, e.g. [6, 7]. Presently, there are still reports of further improvements to the topology and innovations in the design towards increased efficiency, e.g. for photovoltaic applications [8, 9]. At the core of most switch-mode converters lies a circuit consisting of a switch, a diode and an inductor. A simple buck converter circuit for step-down voltage conversion is shown in Fig. 3.2.

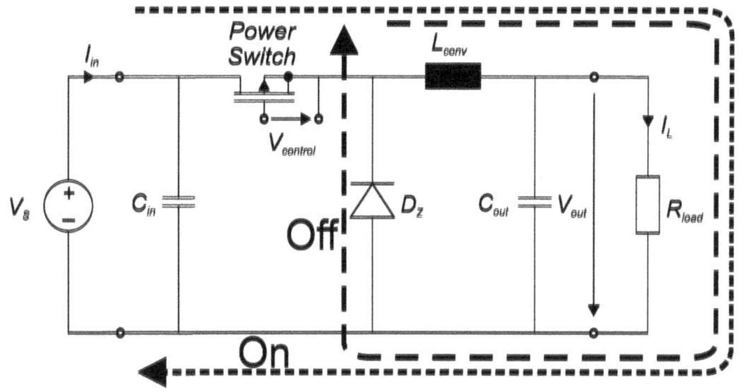

Fig. 3.2 Circuit topology of buck converter circuit for voltage reduction and current paths in on- and off-state of the power switch, respectively

During the charging phase, the semiconductor switch is closed and the inductor current increases while more electrical energy gets stored in its magnetic field. When the semiconductor switch is opened during the discharge phase, the inductor current continues flowing through the freewheeling diode and the inductor current decreases while electrical energy is removed from the system. If the switching frequency and duty cycle of the semiconductor switch are precisely controlled, any voltage below the supply voltage can be applied to the load on the secondary side. With increasing switching frequencies, the energy portions that need to be transferred per switching cycle become smaller and the inductance can be reduced, i.e. a smaller inductor can be used. However, parasitic inductances in the circuit and dynamic power losses in the semiconductor devices limit the switching frequencies that can be applied. Extensive literature on modelling pulse-width modulated switch-mode circuits is available, e.g. a model of a pulse-width modulated switch for both continuous and discontinuous modes of operation [10].

3.1.2 *Benefits and Drawbacks of Switch-Mode Conversion*

The main reason for the success of pulse-width controlled switch-mode converters with active power factor correction is a power factor close to unity, i.e. the energy flowing into the circuit from the primary side is (almost) completely transferred to the secondary side. This maximizes the amount of energy that can be, for instance, drawn from an office wall plug [3]. Moreover, the total harmonic distortion of such a circuit is minimized. Consequently, passive networks at the input of the converter which are bulky and heavy can be avoided [7].

The use of a switch-mode based circuit also allows for efficient energy conversion beyond 90 %, i.e. the power losses incurred during operation can drop below 10 %. Ideally, the power switches operate either at low voltages (on-state) or at low currents (off-state) and little power is consumed. Consequently, the thermal power dissipation is much smaller than in fixed voltage regulators. It is noteworthy that for high switching frequencies the power losses during switching from on- to off-state and vice versa can be significant. This will be discussed in Chap. 4 in detail.

Because the input voltage is typically rectified at the input of the converter, high tolerances regarding the grid frequency can be achieved. Moreover, regulation of the switching frequency allows for a large variance in input voltages. Typical switch-mode circuits for operation of the secondary side at low DC voltage can be connected to both the European (230 V AC, 50 Hz) and the American (110 V AC, 60 Hz) Mains power.

In comparison to voltage transformation (using a transformer and subsequent DC rectification), the conversion solution using a switch-mode circuit achieves much smaller weight and form factors. The same consideration also holds true in comparison to fixed voltage regulators which require larger transistors and a heat sink due to the high average power dissipation [11]. Finally, using a suitable boost converter topology, generation of voltages above supply voltage is also possible.

Among the disadvantages are the higher part count and the added complexity from power factor correction because additional control circuitry is necessary aside of the power semiconductor devices (switches and diodes). Moreover, electromagnetic interference will be generated in the converter due to the hard switching operation at higher frequencies (typically in the kHz range), and suitable low pass filters need to be added to the circuit to avoid this.

Since the energy is not constantly provided, a buffering capacitor has to be employed on the secondary side to minimize the voltage ripple. Increasing switching frequency will reduce the voltage ripple and relax the size requirement for the filter capacitor [12].

3.1.3 Direct-Current Conversion and Solid-State Rectification: Buck and Boost Converters

The conversion of the voltage level at direct current can be performed by different circuit topologies for buck and boost converters. Some relevant topologies without galvanic insulation are shown in Fig. 3.3 to further illustrate the concept of switch-mode conversion.

Beside the buck converter (Fig. 3.3a) which can be used to reduce the voltage level below the input voltage of the circuit, a boost converter (Fig. 3.3b) is capable of generating voltages beyond the input voltage of the circuit. The boost converter charges the inductance using the power transistor in the first phase. Switching the transistor off in the second phase charges the output capacitor—and thus the output voltage—beyond the input voltage level.

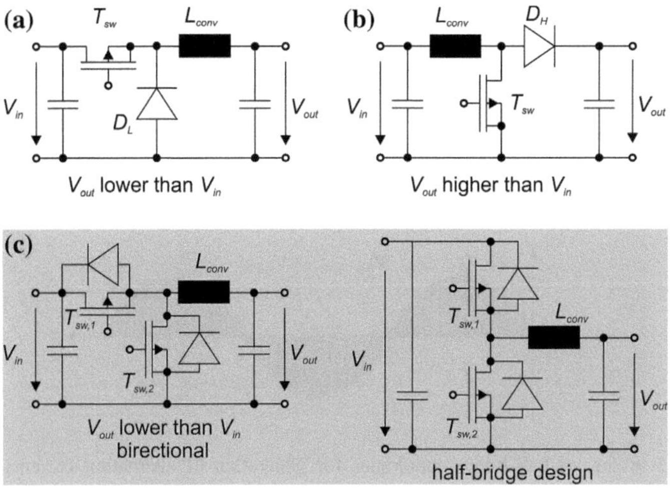

Fig. 3.3 Overview of switch-mode topologies with power factor correction

Additional converter topologies with galvanic isolation between primary and secondary side are presented in literature [4]. Generally, these concepts include transformers to realize galvanic isolation.

3.1.4 Power Inverters: Half and Full Bridge Topologies

When the buck converter is extended by a second switch in parallel to the diode D_L as depicted in (Fig. 3.3c), the energy stored in the output capacitor can be transferred to the inductance by closing power switch $T_{sw,2}$. When this power switch is opened, the current through the inductance will continue to flow through the additional free-wheeling diode parallel to transistor $T_{sw,1}$ into the input capacitor. This operation mode constitutes to a boost converter from the right side to the left side of the circuit. Therefore, bidirectional operation of this converter can be achieved. Both diodes can be implemented as body diode of a power MOSFET or by additional external diodes to improve switching performance, e.g. using unipolar diodes or bipolar diodes optimized for fast switching [13].

Power electronic circuits for the operation of AC motors at variable frequencies for single or three phase operation are also known. The most commonly used circuits employ half-bridge topologies like the one shown in Fig. 3.4. By choosing a pulse-width modulated switching scheme of the six power switches, a sine waveform which is required for operation of an asynchronous motor with torque control can be provided. This bidirectional converter concept also enables recuperation or electric break functionality in electric motor drives. Extending this

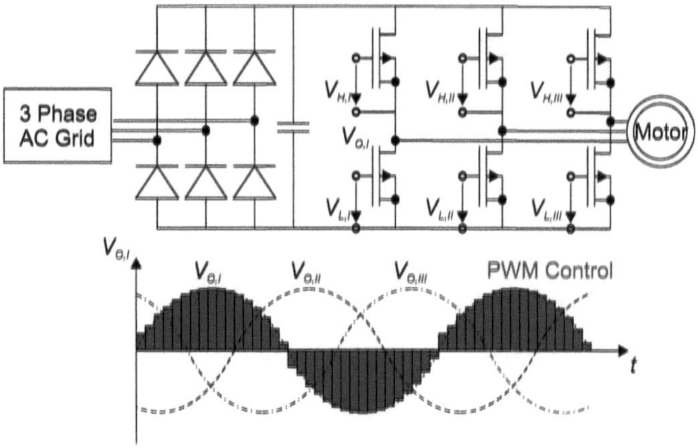

Fig. 3.4 Overview of half-bridge topologies for generation of alternating currents at variable frequencies

topology to a full bridge (H-bridge) scheme even allows for bidirectional buck-boost conversion offering full control regarding voltage level and frequency at the secondary side of the power converter.

Integration of the power transistors and the diodes into a monolithic integrated circuit enables a minimization of parasitic inductances. Thereby, over-voltages and current surges are reduced compared to a discrete assembly using bond wires. Additionally, extending the integration to gate drivers and logic allows for the realization of a robust IC including features like over- and under-voltage protection, short circuit or no-load detection. Again, the reduction of bond wires and solder joints for the whole circuit also increases reliability.

3.2 Circuits for RF Amplifiers

By itself, the information that is transmitted by telecommunication equipment as of today is digital. Nevertheless, the signals that are routed through the RF amplifier circuit have to be considered analogue due to the complexity of modern modulation topologies which include the use of multiple channels and phases in the broadcasted signal. The amplification of analogue signals necessitates a circuit technology that offers high linearity. A digital switching scheme as described in the previous section for converters was not suitable when vacuum tubes and later on bipolar power transistors became available. Amplification circuits for analogue signals are known for almost one century [14, 15]. And even the most basic topologies are still in use today. The fundamental topology of a class A amplifier circuit depicted in Fig. 3.5 consists of a single inverting stage.

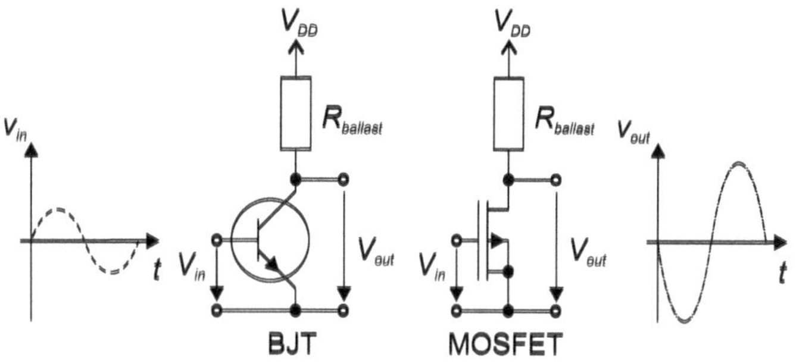

Fig. 3.5 Fundamental amplifier circuit and ideal output signal for a sinusoidal small-signal excitation from the operating point

Assuming an ideal switch, the output signal is an amplified reproduction of the input signal. The switch exhibits a cut-off when the input voltage is below the threshold voltage. The output current increases linearly with input voltage, and the output current is independent of the output voltage. Only at low output voltages (below the knee voltage) a rapid drop to zero current is observed.

3.2.1 Amplifier Fundamentals

For RF amplifiers, a number of distinct properties to enable the maximum system performance are of importance. A high penetration range of the broadcasting system, reducing the number of base stations per square kilometre, requires high output power. In order to maximize the power of the amplified signal, high power efficiency is desirable. This goes hand in hand with the reduction of power losses that also need to be dissipated. Base station equipment requires power switches with output powers of over 100 W [16]. Moreover, handheld devices, which are powered by batteries, achieve longer *talk time* if the RF amplifier is more efficient. Transmission of the information from the mobile phone to the base station requires 2 W, being one of its most energy-demanding functionalities. The power demand of RF amplifiers in mobile phones is very diverse across different countries and geographical areas, e.g. due to legislation [17]. The components of an RF amplifier and the associated current flow are schematically presented in Fig. 3.6.

The output matching network consists of a harmonic termination which shorts out harmonic frequencies and a fundamental matching network. The fundamental matching network transforms the load line resistance of the RF amplifier to the impedance level of the system [18].

Drain Efficiency By comparing the supplied power consumed by the circuit (DC bias) P_{DC} with the output power of the amplified signal P_{out}, the drain efficiency η_D can be calculated.

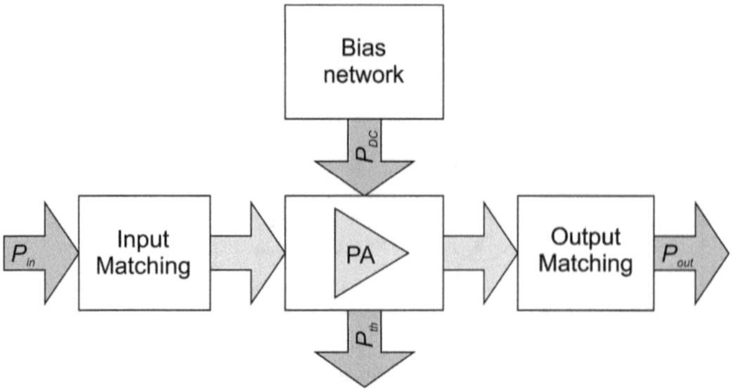

Fig. 3.6 Power distribution in RF amplifiers

$$\eta_D = \frac{P_{out}}{P_{DC}} \tag{3.1}$$

Since the termination resistance of the system is typically 50 Ω, the ideal operating voltage for the RF amplifier which has to be provided by the bias network at a given power level is given by

$$P_{out} = \frac{V_{DC}^2}{R_{Load}} = \frac{I_{DC}.V_{DC}}{2}, \tag{3.2}$$

which is valid when the input power can be neglected (high level of amplification) and a sinusoidal waveform of the input signal is assumed. The average supply power of an ideal amplifier circuit in contrast equals to

$$P_{DC} = I_{DC} \cdot V_{DC}, \tag{3.3}$$

yielding an output efficiency of 50 % for a class A amplifier using a sinusoidal waveform. In reality, it will be significantly lower due to the power of the input signal and device cut-off at the knee voltage.

When the level of amplification is low (e.g. the input signal has already been amplified in a pre-amplification stage), the power of the input signal P_{in} represents a significant contribution to the output power of the amplifier and should be considered.

Power Gain The ratio of the output power P_{out} to the input power P_{in} of the amplified signal is defined as power gain G_p

$$G_p = \frac{P_{out}}{P_{in}} \tag{3.4}$$

A concatenation of several amplifier stages as multi-stage MMICs can be used to increase the power gain.

Power Added Efficiency The efficiency of an RF amplifier including the energy provided by the input is given by the power added efficiency (*PAE*)

$$PAE = \frac{P_{out} - P_{in}}{P_{DC}}, \tag{3.5}$$

further reducing the energy efficiency compared to an ideal circuit. The reduction becomes particularly apparent for low power gains, i.e. when the output signal is not much stronger than the input signal. The *PAE* can be expressed as

$$PAE = \eta_D \left(1 - \frac{1}{G}\right). \tag{3.6}$$

For a power gain of 100, the *PAE* is within 1 % of the drain efficiency [19]. For lower gain, a significant reduction of *PAE* starts to occur. Increasing the power gain beyond 100 does not significantly increase energy efficiency but implies additional effort in circuit design. Therefore, a power gain of 100 (20 dB) is usually desired for each stage of a RF amplifiers.

Linearity A high linearity of the RF amplifier minimizes distortion of the signal. In code-division or orthogonal frequency-division modulation access schemes used for UMTS and LTE communication equipment, the fulfilment of linearity specifications like Adjacent Channel Power and Alternate Channel Power is of utmost importance [20]. Based on this fulfilment, other properties like *PAE* and power density are then optimized. Figure 3.7 represents the correlation between input and output power for an ideal and a real (saturation of power transistor) class A amplifier.

Ideally, the output power increases linearly with higher input power and the power gain is constant. In reality, at a sufficiently high input power, the circuit saturates, the power gain reduces and the input signal is compressed. Typically, a compression of 1 dB is still acceptable for modern modulation schemes in telecommunication [21]. The request for higher linearity reduced the maximum output power of a given amplifier circuit.

Dynamic Range When multiple channels with different frequencies are asserted, the transmitted output signal is the sum of the amplified input signals (channels). A high dynamic range of the RF amplifier enables the utilization of a large number of channels. The output signal of a multi-channel transmission over time is depicted in Fig. 3.8.

Peak-to-Average Ratio The ratio between the highest output signal and the average signal is known as the peak-to-average ratio (*PAR*). It is defined as

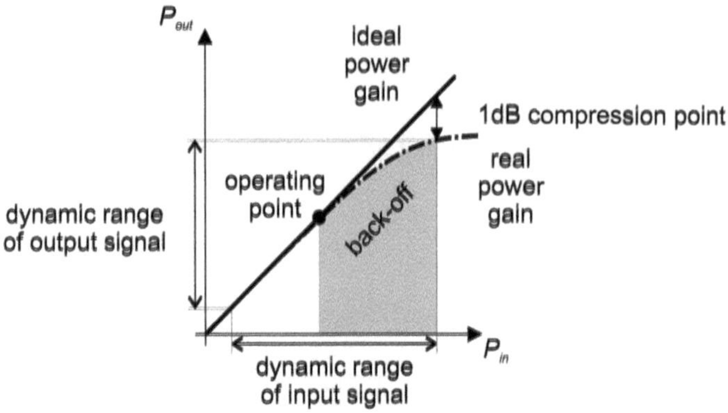

Fig. 3.7 Input and output power of RF amplifier; comparison between ideal and real circuit characteristics in a class A amplifier

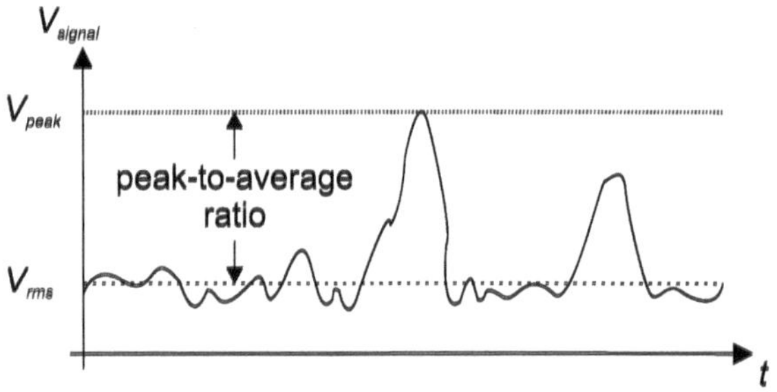

Fig. 3.8 Peak-to-average ratio: Result of channel intermodulation as the sum of signals with different frequencies

$$PAR = \frac{V_{peak}}{V_{rms}}. \tag{3.7}$$

Obviously, successful reconstruction of the transmitted signal requires that the signal is not compressed beyond the 1 dB point even when peak signals occur.

Back-Off Therefore, the amplifier circuit must be operated in back-off: The *PAR* defines the distance of the operating point for a single channel signal below the 1 dB compression point (see Fig. 3.7).

Bandwidth The bandwidth describes the frequency range available for each transmission channel in frequency-division multiple access protocols. While it is not a characteristic of the RF amplifier itself, it should be noted that a larger bandwidth reduces the linearity constraint because more signal distortion can be accepted [22]. A circuit with a high power density enables the design of smaller and cheaper RF amplifiers. Here, the power density is limited by the semiconductor switches used in the circuit. The same is true for the output power with respect to Power Added Efficiency at a given frequency.

Table 3.1 Summary of RF amplifier circuit requirements for handhelds and base stations

Circuit property	Importance	
	Handheld	Base station
Power added efficiency	High	Low
Output power density	High	High
Linearity (time and frequency-division multiple access), e.g. GSM	High	High
Linearity (orthogonal frequency and code-division multiple access), e.g. UMTS, WiMax, LTE	Very high	Very high

Gain-Bandwidth Product The gain-bandwidth product describes the relationship between the current gain of the amplifier circuit and its operation frequency. Under typical operating conditions the product remains constant, i.e. if the switching frequency of an RF amplifier is lowered, the bandwidth increases. Conversely, if a sufficient current gain is required at a high operating frequency, the amplifier must exhibit a large gain-bandwidth product. This makes the gain-bandwidth product a decent Figure-of-Merit for comparing different amplifier circuits with respect to their high frequency performance. It can also be converted to the transition frequency of the amplifying semiconductor device as will be shown in Chap. 4.

Stability of Output Signals The circuit must also exhibit sufficient stability to ensure proper operation for the specified lifetime of the product. Again, most of the stability issues arise from the semiconductor device itself or the associated connection and assembly technology. They will be covered in details in Chap. 4. There is a trade-off between the power density, output power and the degradation over time. The requirements for circuits in RF amplifiers are summarized in Table 3.1.

3.2.2 Amplifier Classes

As previously mentioned, a class A amplifier cannot exceed a drain efficiency of 50 %. For solutions where a higher efficiency is required, a different amplifier topology needs to be exploited. Due to different performance requirements, a wide range of amplifier classes have been developed [18]. Two prominent amplifier classes are compared in Fig. 3.9.

The main different between these topologies is the location of the operating point. The class A amplifier operates at the centre of the load line so that both polarities of a sine wave can be amplified. In contrast, a class B amplifier exhibits an operating point right at the cut-off point. This minimizes the power losses, because the circuit is consuming little power in standby mode. However, a second power switch is necessary to amplify the negative half-wave (push-pull stage). Also, due to the strong reduction of power gain below the knee voltage, strong distortion may occur for very small signals. A trade-off between these two circuit

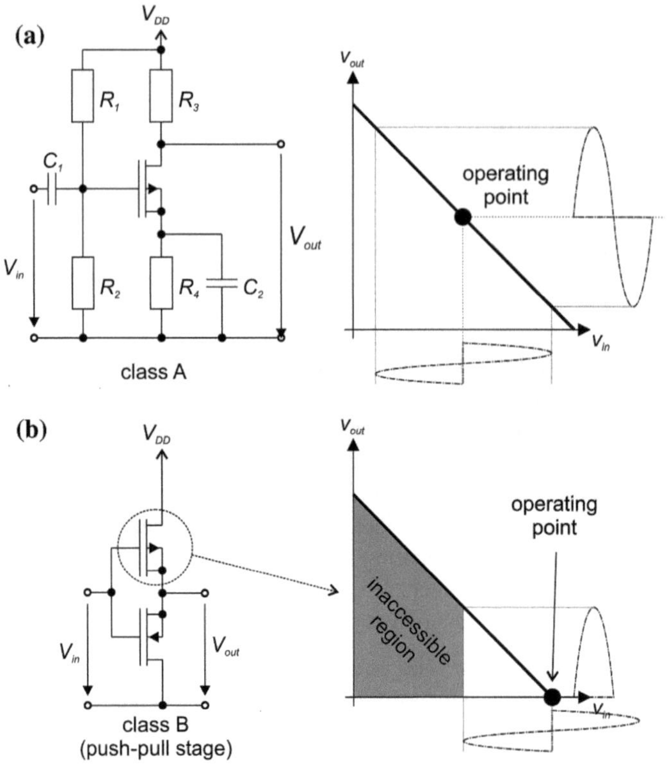

Fig. 3.9 Overview of basic RF amplifiers: **a** class A and **b** class B with their respective load lines

topologies is the class AB amplifier which sacrifices some power efficiency for the sake of linearity: Here, the operating point is located right above the knee voltage and distortion can be minimized.

Recently, amplifier classes with theoretical power efficiency of 100 % have gained additional interested: For audio amplifiers, where the signal is limited to frequencies below 20 kHz, the class D amplifier has been introduced [23]. In contrast to the previously mentioned class A and B amplifiers where high linearity is achieved by operation of the transistors in a linear portion of the load line, the class D type relies on a switch-mode concept very similar to switch-mode amplifiers where the amplified signal is "reconstructed" by high frequency switching [24]. This concept can readily be applied in audio amplifiers where signal frequencies are in the upper kHz range and adequate sampling frequencies two or three decades higher are feasible [18].

In order to use this amplifier topology at RF frequencies, the class D amplifier can be extended to the so-called class S amplifier [25] as shown in Fig. 3.10.

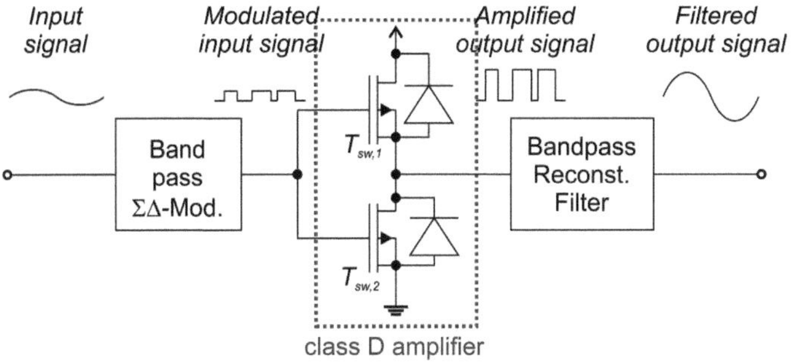

Fig. 3.10 Class S amplifier for RF signal amplification using a switch-mode based topology, adopted from [25]

Without going into detail about the operating principle, it is evident that the circuit technology used in a class S amplifier resembles a switch-mode circuit routinely used in energy conversion, i.e. in switch-mode power supplies. Power semiconductor devices used in this amplifier are operated in linear operation mode which is in contrast to conventional amplifiers (e.g. Class A, B) where the saturation region is used, and this enables a high drain efficiency.

3.3 Application-Specific Requirements for Power Transistors

From application and circuit requirements, the desirable properties of power transistors can be extracted. The following summary lists device requirements and derives implications for the design of power semiconductor devices.

Energy Efficiency and Frequency The energy efficiency of a power electronic system is limited by the properties of the power semiconductor devices and the circuit topology. Power dissipation arises from both static losses and dynamic losses. On one hand, switch-mode applications are mainly dealing with static losses and dynamic losses incurring from the transition between on- and off-state. On the other hand, power losses in RF amplifiers are strongly affected by dynamic losses incurred from charging and discharging the parasitic device capacitances.

Power Density and Thermal Losses The demand for high power densities arising from the economic point of view can be satisfied by semiconductor devices. By passing high currents over the device a smaller chip size can be used. Theoretically, power devices with little static losses can be fabricated, but they would turn out too expensive for mass-market applications. Also, when operating at high frequencies, a large chip also yields large parasitic capacitances that need to be charged and discharged. Both considerations give rise to the demand for small chips with high

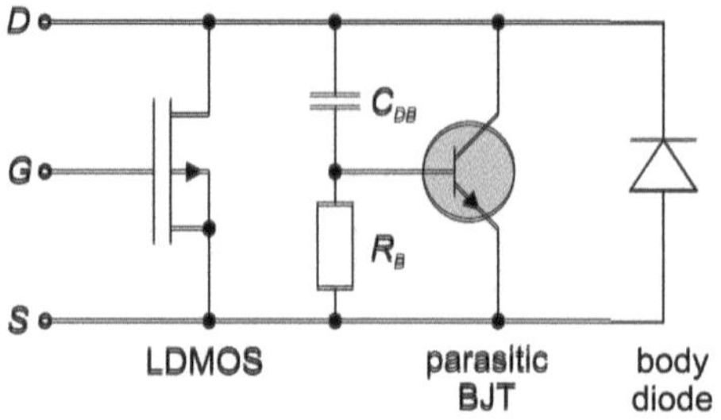

Fig. 3.11 Equivalent circuit of LDMOS transistor with parasitic bipolar junction transistor and body diode

power densities. This in turn requires good cooling of the power devices to remove the high power losses dissipated locally in the system.

Reliability and Ruggedness Considerations Exceeding the safe operating area of power semiconductor devices may temporarily yield superior device performance —and it is a valid option in some applications where long lifetime is not crucial— but it will result in severe device degradation and premature device failure. Examples include the injection of hot carriers and the degradation of the gate dielectric at excessive gate voltage both shifting the knee voltage of RF amplifiers. Given that most power electronic applications are based on digital switching of the power device (On-Off), the device degradation is less critical compared to an analogue switch used in RF amplifiers.

Other critical operating conditions which can arise from high currents, high voltages or high voltage changes are associated with dynamic properties. An example for premature device failure due to high drain currents or a rapid change in drain current at high drain voltages is the turn-on of the parasitic bipolar junction transistor in an LDMOS transistor as shown in Fig. 3.11.

In this event, the device current exceeds the specification for which the device has been designed. At high power densities, local device heating leads to an increase in the bipolar current, thermally inducing the bipolar turn-on and corresponding device failure [26]. Proper design of power semiconductors must be employed to achieve the required reliability at the highest possible performance.

Availability of Device Models The availability of suitable device models is an aspect that should also be mentioned when discussing requirements for power semiconductor devices. For a precise and reliable circuit simulation, adequate device models including temperature dependence are necessary. Circuit simulation has become a key method of reducing time-to-market and to lower the development costs. Clearly, semiconductor devices that cannot be reliably modelled can hardly be integrated into an application that relies on precise control of device properties.

Table 3.2 Summary of semiconductor device properties and their importance for power electronic circuits and RF amplifiers

Power device property	Relevancy to circuit/system	
	Power electronics	RF amplifiers
Energy efficiency	High	Low to average
Power density	High	High
Linearity and gain	Low	Very high
Reliability	High	High
Manufacturing cost	High	High

3.4 Summary of Power Semiconductor Device Requirements

Due to the wide range of applications, a clear ranking of the different requirements does not exist. Too broad are the individual demands in different markets. While automotive applications require extremely rugged and reliable devices, consumer products are primarily cost driven. This is not to say that the cost pressure in the automotive market is low either. Nevertheless, some trends regarding device requirements can be derived. A rough comparison between power device properties for power electronic and RF amplifier circuits is presented in Table 3.2.

Requirements for similar circuit topologies vary between different applications, however, making a universal guidance near to impossible.

References

1. M.D. Perry, M. Gerard, Terawatt to petawatt subpicosecond lasers. Science **264**(5161), 917–924 (1994)
2. R.J. Widlar, New developments in IC voltage regulators. IEEE J. Solid-State Circuits **SC-6**(1), 2–7 (1971)
3. M.F. Schlecht, Harmonic-free utility/DC power conditioning interfaces. IEEE Trans. Power Electron. **1**(4), 231–239 (1986)
4. A.I. Pressman, K. Billings, T. Morey, *Switching Power Supply Design*, 3rd edn. (McGraw-Hill Education, Columbus, 2009)
5. T. Matsuhashi, Y. Amemiya, Thyristor power control with less third harmonic component. J. IEE (Jpn) **90**, 1621–1627 (1970)
6. T. Kataoka, K. Mizumachi, S. Miyairi, A pulsewidth controlled AC-to-DC converter to improve power factor and waveform of AC line current. IEEE Trans. Ind. Appl. **15**(6), 670–675 (1979)
7. Z. Yang, P.C. Sen, Recent developments in high power factor switch-mode converters, in *IEEE Canadian Conference on Electrical and Computer Engineering* (Waterloo, Canada, 1998)
8. B. Burger, D. Kranzer, Extreme high efficiency PV-power converters, in *European Conference on Power Electronics and Applications (EPE)* (Barcelona, Spain, 2009)

9. T. Kerekes, R. Teodorescu, P. Rodriguez, G. Vazquez, E. Aldabas, A new high-efficiency single-phase transformerless PV inverter topology. IEEE Trans. Ind. Electron. **58**(1), 184–191 (2011)
10. V. Vorperian, Simplified analysis of PWM converters using model of PWM switch part I: continuous conduction mode. IEEE Trans. Aerospace Electron. Syst. **26**(3), 490–496 (1990)
11. K. Billings, T. Morey, *Switchmode Power Supply Handbook*, 3rd edn. (McGraw-Hill, New York, 2010)
12. I.W. Clelland, R.A. Price, Recent advances in capacitor technology with application to high frequency power electronics and voltage conversion, in *Applied Power Electronics Conference and Exposition* (Dallas, USA, 1999)
13. L. Lorenz, G. Deboy, I. Zverev, Matched pair of CoolMOS transistor with SiC-Schottky diode —advantages in applications. IEEE Trans. Ind. Appl. **40**(5), 1265–1272 (2004)
14. J.H. Morecroft, H.T. Friis, The vacuum tubes as a generator of alternating-current power. Trans. Am. Inst. Electr. Eng. (AIEE) **38**, 1415–1444 (1919)
15. D.C. Prince, Vacuum tubes as power oscillators—part I, Proc. Inst. Radio Eng. **11**(3), 275–313, 1923
16. K. Joshin, T. Kikkawa, H. Hayashi, T. Maniwa, S. Yokokawa, M. Yokoyama, N. Adachi, M. Takikawa, A 174 W high-efficiency GaN HEMT power amplifier for W-CDMA base station applications, in *IEEE Electron Devices Meeting* (Washington, USA, 2003)
17. S. Lönn, U. Forssen, P. Vecchia, A. Ahlbom, M. Feychting, Output power levels from mobile phones in different geographical areas; implications for exposuer assessment. Occup. Environ. Med. **61**, 769–772 (2004)
18. S. Cripps, in *Amplifier classes, A to S*, ed. by John L. B. Walker. Handbook of RF and Microwave Power Amplifiers (Cambridge University Press, Cambridge, 2012), pp. 159–187
19. W. Burger, C. Dragon, in *Silicon LDMOS and VDMOS transistors—Physics, design and technology*, ed. by John L. B. Walker. Handbook of RF and Microwave Power Amplifiers (Cambridge University Press, New York, 2012), pp. 1–39
20. C.E. Weitzel, RF power devices for wireless communications, in *IEEE MTTS International Microwave Symposium* (Seattle, USA, 2002)
21. A. Springer, T. Frauscher, B. Adler, D. Pimingsdorfer, R. Weigel, Impact of nonlinear amplifiers on the UMTS system, in *IEEE International Symposium on Spread-Spectrum Technology and Application* (New Jersey, USA, 2000)
22. D.P. Myer, in *Practical HF/VHF/UHF RF power amplifier realization*, ed. by John L. B. Walker. Handbook of RF and Microwave Power Amplifiers (Cambridge University Press, Cambridge, 2012), pp. 232–283
23. F.H. Raab, Get broadband, dual-mode operatin with this FET power amplifier, Electr. Design News. **23**, 117–124 (1978)
24. H.L. Krauss, C.W. Bostian, F.H. Raab, *Solid-State Radio Engineering* (Wiley, New York, 1980)
25. T. Johnson, S.P. Stapleton, RF class-D amplification with bandpass sigma-delta modulator drive signals. IEEE Trans. Circuits Syst. I **53**(12), 2507–2520 (2006)
26. S.J.C.H. Theeuwen, J.A. de Boet, V.J. Bloem, W.J.A.M. Sneijers, LDMOS ruggedness reliability. Microwave J. **5**, 96–104 (2009)

Chapter 4
Power Semiconductor Devices in Power Electronic Applications

This chapter introduced topologies for power semiconductor devices and explains the suitability of lateral power MOSFETs for implementation in integrated circuits. It begins with a thorough definition of power semiconductor devices in general and lateral power transistors in particular because these terms are ambiguously used in different traits of microelectronics and microwave technologies. The fundamental requirements for power semiconductor devices are also discussed. Next, device topologies regularly found in state-of-the-art power electronic and RF applications are reviewed. This includes vertical power MOSFETs, stand-alone lateral RF MOSFETs and lateral power MOSFETs. Then, the trade-offs between different aspects of these power semiconductor devices are reviewed, and a summary of *Figures-of-Merit* (FOM) suitable for comparison of different power semiconductor devices is presented. These FOMS are derived from the application point-of-view following the requirements set forth in Chap. 2. They will be used throughout the textbook to compare different device technologies regarding their applicability in relevant circuit topologies. For the applications in Chap. 2, combination with logic and memory devices, reliability and robustness demands and small size and low weight of the overall circuit are key elements. This emphasizes the advantages of lateral power MOSFETs in integrated circuits. However, limitations and drawbacks of lateral power transistors need to be considered as well. In particular, the combination with logic and memory devices comes with a cost to process freedom, and complicates implementation of junction termination concepts. Moreover, unipolar charge transport and field enhancement at the device surface result in lower performance compared to stand-alone power devices.

4.1 Introduction to Power Semiconductor Devices

The mostsuccessful semiconductor devices of today are the p-channel and n-channel MOSFETs constituting to CMOS logic. Without a doubt, these devices have contributed to the rise of information technologies as they can be found in central processing units, graphic accelerators, digital sound processing and in both

T. Erlbacher, *Lateral Power Transistors in Integrated Circuits*,
Power Systems, DOI 10.1007/978-3-319-00500-3_4

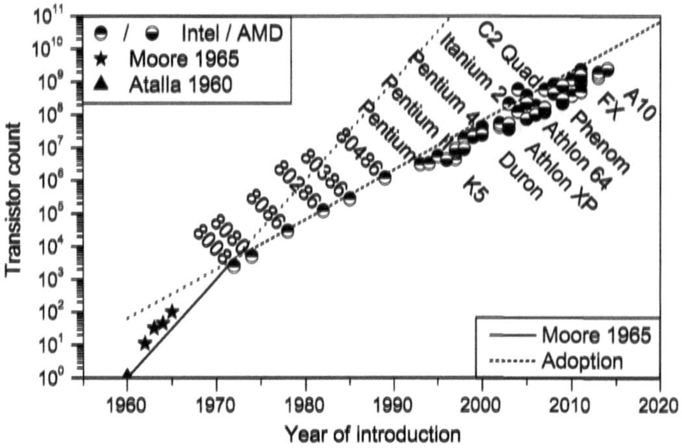

Fig. 4.1 Component count per integrated circuit for microprocessors and initial prediction by G. Moore, data from [1] extended by Intel and AMD CPU release dates

volatile dynamic random access memory and non-volatile electronically erasable programmable random access memory. These components are the basic building blocks of computers and telecommunication equipment. By following Moore's Law [1], which predicted a doubling of devices per integrated circuit every 2 years (from the economic point of view), the integration density of these semiconductor devices has rapidly increased over the last 60 years. This development is shown in Fig. 4.1 depicting the release dates of Intel and AMD microprocessors.

In 2013 Intel has introduced a microprocessor fabricated in 22 nm technology capable of running at 4.0 GHz with 1.86 billion transistors located on one chip of 177 μm^2 size.

However, these MOS-based logic transistors are not capable of withstanding operating voltages beyond 5–10 V. Moreover, this massive shrinking of device size has further reduced operating voltages of CPUs from 5 to 3.3, 1.8 and 1.3 V [2]. With the application of a constant electric field approach, it is apparent that future logic devices will use even lower operating voltages.

Given that computers and servers are usually connected to the alternating current grid infrastructure, voltages from 110 to 230 V need to be controlled and converted to the low operating voltages which can be sourced to the logic devices. In modern computers, there are several DC voltage levels for peripheral components like hard disk drives, USB ports and fans which run at 5 and 12 V, respectively. Figure 4.2 shows a schematic summary of these voltage levels in a typical desktop computer.

Apparently, there is a significant demand for semiconductor devices that are capable of handling operating voltages beyond the logic level, enabling the operation of the logic devices from the grid or another high voltage supply. The MOS-based logic device is failing at higher drain-source voltages because either the electric field in the channel region exceeds the critical value and avalanche breakdown occurs or punch-through breakdown occurs due to the low doping of the

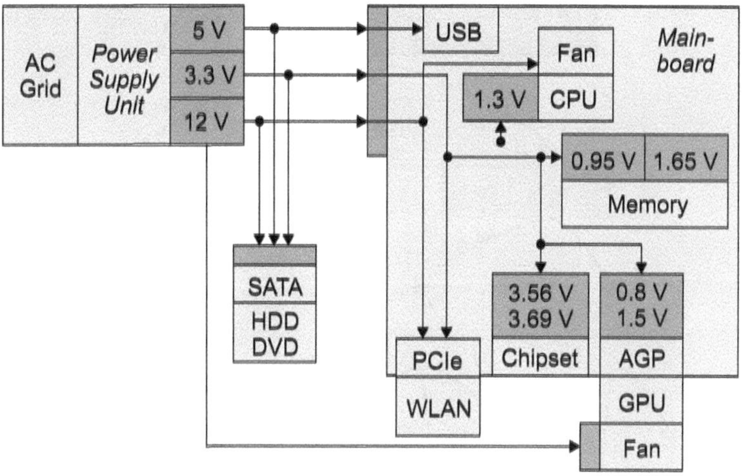

Fig. 4.2 Voltage levels and regulators for components and peripherals of a personal computer

channel region. Additionally, the thin gate dielectric is not capable of withstanding high gate-drain voltages without degradation or even dielectric breakdown.

The approach that has been widely used to increase the voltage rating of semiconductor devices is to provide an extension of the drain region—the *drift region*—which is lightly doped compared to the drain region. The space-charge region extending into this drift region will then carry most of the blocking voltage applied to the drain electrode, relaxing the electric field across the channel. Moreover, with a doping concentration significantly lower than the channel, punch-through conditions can also be prevented.

Regarding a definition of power semiconductor devices, it can be stated that:

> Any semiconductor device that includes dedicated device structures, e.g. a lightly doped drift region, in order to withstand drain-source voltage beyond logic level can be considered a power semiconductor device.

A large number of power semiconductor devices have evolved in the last century with the huge majority of technological development on silicon. The most prominent devices based on silicon technology are listed in Fig. 4.3.

Among these, the unipolar LDMOS transistors have been extensively used in integrated circuits and radio-frequency amplifiers. These devices will be discussed throughout the remainder of this book in detail. A lateral drain-extended MOS transistor was first introduced by Sigg et al. in 1972 [3]. Its cross-section is schematically shown in Fig. 4.4.

The device was named DMOS transistor due to the simultaneous double diffusion of source and channel dopants that was required to fabricate a short channel device for high frequency operation. The p-well region containing the MOS channel is formed due to the faster diffusion of these p-dopants (boron) compared to the dopants used for the n^+-source region (arsenic or phosphorous). Despite the

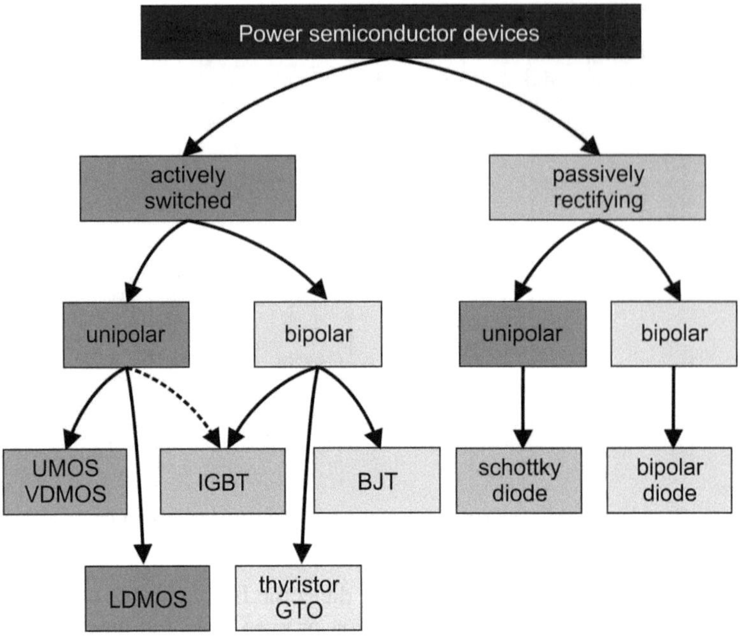

Fig. 4.3 Summary of power semiconductor devices fabricated in silicon

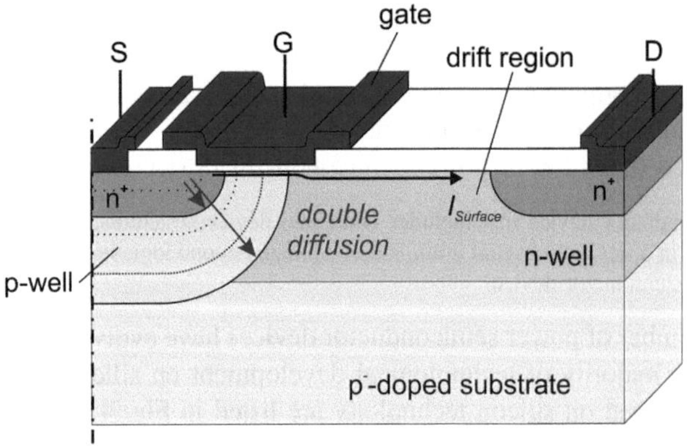

Fig. 4.4 Cross-section of a DMOS transistor with lightly n-doped drift region, adopted from [3]

availability of ion implantation and high resolution lithography to manufacture wells, the term DMOS has also been adopted for power semiconductors that do not rely on this double diffusion concepts nowadays like the VDMOS and LDMOS transistors for power electronic applications. These devices exhibit a channel region similar to the double-diffusion process.

Depending on the voltage and current ratings and whether integration with logic devices is desired, two different device concepts for MOS-based power semiconductor devices have evolved.

- The vertical power MOSFET is typically employed for high power densities in single device packages or power modules.
- The lateral power MOSFET is used when integration with logic devices or high operating frequencies as in radio-frequency power amplifiers are required.

In the context of this textbook the terms lateral and vertical define the direction in which the drain-source voltage is blocked within the semiconducting material and the drain current is flowing through the drift region.

This means that even if the drain contact is located at the backside of the device the presence of a lateral drift region makes it a lateral MOSFET device. This is both true for a typical RF LDMOS and a LDMOS transistor in a smart-power IC. "Quasi-vertical" transistors exhibit vertical current flow through the drift region but have source and drain contacts on the same side of the semiconductor. Also, this implies that both VDMOS and UMOS transistors are vertical power devices, even though the VDMOS has a lateral channel and the channel in the UMOS is vertical.

Among the MOS-based power transistors, the vertical device topologies have received a lot of attention, because they are able to carry larger current densities than their lateral companions. The capability to inject the current from one side of the chip and to remove it on the other side provides a straight forward approach for the metallization patterns. A cross-section of a typical VDMOS transistor is presented in Fig. 4.5.

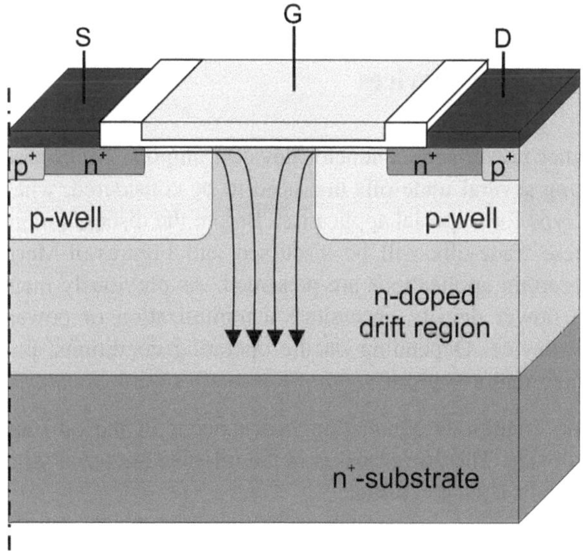

Fig. 4.5 Cross section of a VDMOS transistor with a lateral channel region and vertical drift region

Here, the current flows laterally from the source through the channel region into the n-doped drift region. From there it can spread under the cell to pass vertically through the drift region where it is collected at the drain contact. Assembly technologies using direct bonded copper substrates or with e.g. TO-220 packages are available to provide solutions for highly integrated power inverters and buck/boost converters. An increase in current density can be facilitated by integration of trench gates, i.e. UMOS transistors.

Further advances in vertical power transistor technology have also led to the development of superjunction devices with very low on-state resistances [4]. The application of this charge-compensation concept to LDMOS transistors is also investigated in Chap. 6.

The key benefit of using semiconductor technology to fabricate power semiconductor devices is its ability to achieve the circuit and system demands with high reproducibility and reliability under the desired operating conditions. This also includes the availability of these devices at economically feasible fabrication costs and quantities. Increasing wafer sizes and yield as well as the reduction of process variations have added to this development.

A full introduction to semiconductor device physics is omitted in this textbook. For the understanding of this textbook, relevant physical models include doping of semiconductors, diffusion and drift, (temperature and doping dependent) bulk mobility, field-dependent channel mobility, MOS capacitance, impact ionisation and avalanche breakdown. Several excellent textbooks are available dealing with these topics for silicon semiconductors in general [5] and power semiconductor devices [6] and MOS devices [7] in particular.

4.2 Trade-Offs and Figures-of-Merit for Power Semiconductor Devices

Like for all other power semiconductor devices, unipolar MOS-based power transistors are facing several trade-offs that need to be considered, when choosing the proper device type for a special application. Before the discussion on typical device topologies, these trade-offs will be discussed and Figures-of-Merit to judge the suitability for certain applications are presented. As previously mentioned, energy efficiency and power density necessitate a minimization of power losses in the semiconductor device. Depending on the operating conditions, power losses are incurred by three major sources:

- Under static conditions conduction losses occur in the on-state of the semiconductor device. The power losses in the off-state are usually insignificant due to the low drain leakage current.
- Switching losses arise due to finite switching time which causes a current flow while the blocking voltage is still applied across the device.

- Switching losses from charging and discharging the device capacitances can also become significant when a high switching frequency is applied.

These contributions to power losses and Figures-of-merit suitable for judging device performance with respect to these sources are discussed next.

4.2.1 *Static Power Losses: On-State Resistance and Blocking Voltage*

Static power losses result from operation of the power semiconductor device in its on-state. That is, a forward current I_{on} gives rise to a forward voltage drop V_{on} within the device.

$$P_{loss,on} = V_{on} \cdot I_{on}. \tag{4.1}$$

Under static conditions, the forward voltage drop can be expressed as an ohmic total device resistance between drain and source $R_{DS,on}$ at the given forward current.

$$P_{loss,on} = R_{DS,on} \cdot I_{on}^2. \tag{4.2}$$

The contributions of the different regions within the device to the total resistance are summarized in Fig. 4.6 for an n-channel LDMOS device.

For p-channel devices similar considerations hold true. However, the hole mobility in silicon is three times lower than the electron mobility, making n-channel devices the choice for small sized, energy-efficient solutions.

Modern fabrication technologies provide low contact resistances and thick metal lines to keep the resistances of drain and source metallization in check. Significant contributions towards the total resistance arise from the channel and drift regions.

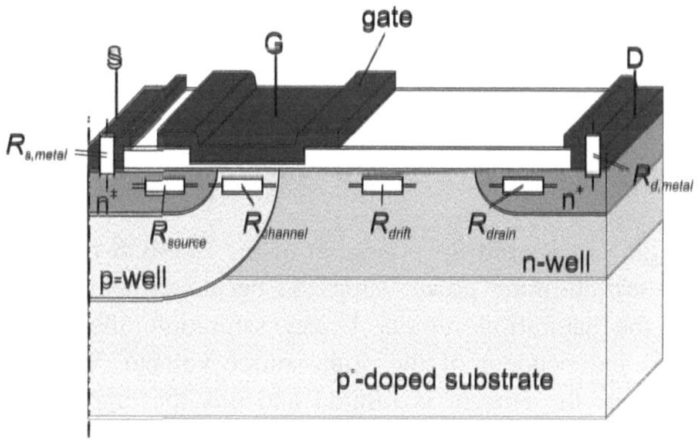

Fig. 4.6 Device resistance contributions within a lateral DMOS transistor

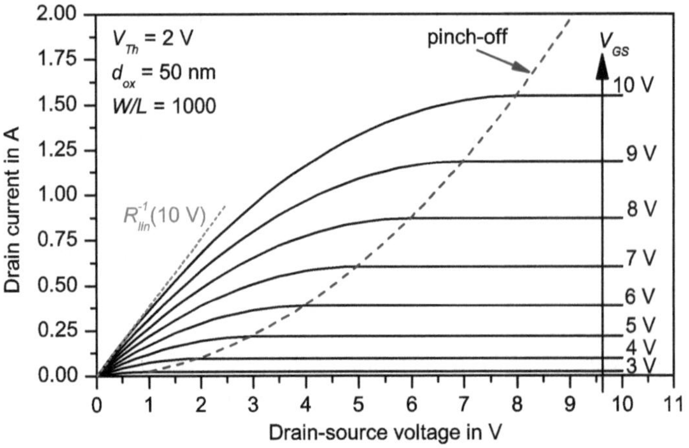

Fig. 4.7 Output characteristics of ideal silicon MOS field effect transistor

The resistance of the channel region can be directly extracted from the output characteristics of a MOS transistor (Fig. 4.7).

In the linear region of operation, the drain voltage is small compared to the gate overdrive (the difference between gate voltage V_{GS} and threshold voltage V_{Th} in the on-state of the device). The drain current is approximated to

$$I_D = \beta(V_{GS} - V_{Th})V_{DS}. \tag{4.3}$$

The transconductance β is calculated from geometric and material constants, namely the inversion channel mobility of electrons $\mu_{n,ch}$ (for an n-channel MOS transistor), the gate capacitance C_{ox}, length of the channel region L_{ch}, and the width W of the device. In silicon, the inversion channel mobility is very close to the bulk mobility for low vertical electric fields. However, when a high gate voltage is applied, the mobility is reduced due to phonon scattering and surface roughness scattering effects [8]. Also, the product of gate capacitance and gate overdrive equals to the amount of channel charge influenced by the gate voltage. From (4.3) the channel resistance in the linear region becomes

$$R_{ch,lin} = \frac{1}{\beta(V_{GS} - V_{Th})} = \frac{L_{ch}}{W \cdot \mu_{n,ch} \cdot C_{ox}(V_{GS} - V_{Th})}. \tag{4.4}$$

The linear region is desired for operation of the LDMOS transistor as a power switch (e.g. for switch-mode power supplies) because it offers a lower channel resistance than the saturation region. Under saturation, the channel resistance becomes (nearly) independent of the drain-source voltage. While this results in higher power losses, the current is limited and unstable operation conditions at excessive drain-source voltages are avoided.

The resistance of the drift region can be readily computed when current spreading effects from the channel into the drift region are neglected. Depending on the average doping concentration N_{Drift} in the drift region, the depth d_{drift} as well as length of the drift region L_{drift} and the width of the device W_{drift}, the resistance for an n-doped drift region is

$$R_{drift} = \rho_{drift} \frac{L_{drift}}{W_{drift} \cdot d_{drift}} = \frac{L_{drift}}{q \cdot \mu_{n,drift} \cdot N_{drift} \cdot W_{drift} \cdot d_{drift}}. \quad (4.5)$$

From an electrical point of view, the total device resistance becomes minimal in the linear mode of operation for the maximum gate voltage, i.e.

$$R_{loss,on} = \left. \frac{V_{DS,lin}}{I_D} \right|_{V_{GS,max}}, \text{with } V_{DS,lin} \ll V_{GS} - V_{Th}. \quad (4.6)$$

This also constitutes to the typical operating point of a power semiconductor device. Comparing the channel and drift region resistances, it becomes apparent that the total device resistance can be reduced by increasing the width of the device —or in other words, by parallelization of a large number of transistor cells. This reduces the current density flowing through the device and in turn the power losses. However, this approach—as discussed later on—also requires a larger chip size which increases the cost of the chip.

In order to compare the suitability and performance of different device technologies, an area independent measure for the total device resistance has been introduced which includes the device width W_{cell} and the device length L_{cell} of an individual cell. This cell constitutes the smallest building block of a power semiconductor device. Based on the channel and drift resistances as the major contributions, the area-independent device resistance $R_{DS,on}$ of an LDMOS transistor equals to

$$R_{DS,on} = \frac{L_{Cell} \cdot L_{ch}}{\mu_{n,ch} \cdot C_{ox}(V_{GS} - V_{Th})} + \frac{L_{Cell} \cdot L_{drift}}{q \cdot \mu_{n,drift} \cdot N_{drift} \cdot d_{drift}} + R_{source} + R_{drain} + R_{metal} \quad (4.7)$$

For typical LDMOS transistors, the first two terms in (4.7) are dominating over the other resistance components. A lower drift region resistance could be achieved by increasing its doping concentration. However, the doping concentration of the drift region also defines the blocking capability of an LDMOS transistor. This can be explained by assuming a simple device structure as shown in Fig. 4.8.

The drain-source voltage will be absorbed in the drift region acting as the lightly-doped region of a pin-diode. Under these circumstances, the highest voltage which the device supports in static condition without failure can be calculated. Avalanche breakdown due to excessive impact ionisation occurs at the breakdown voltage BV when the critical electric field E_{crit} is exceeded [9, 10]. If the drift region is long,

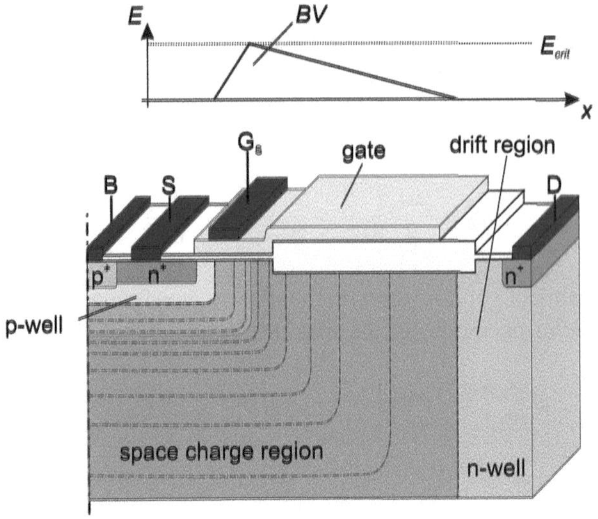

Fig. 4.8 LDMOS transistor with pin-type drift region in non-punch through condition—space charge region and electric field lines are simplified

this will happen before the space charge region fully extends from the p-well/n-drift region junction to the n^+-drain region (non-punch through). Then, the breakdown voltage is given by Poisson's equation to

$$BV = \frac{1}{2} E_{crit} w_{scr,max}. \tag{4.8}$$

In non-punch through condition, the maximum width of the space charge region w_{scr} depends on the maximum electric field (at the pn-junction) according to

$$w_{scr,max} = -\frac{\varepsilon_0 \varepsilon_r}{q N_{drift}} E_{max}, \tag{4.9}$$

resulting in the following expression for the breakdown voltage:

$$BV = \frac{1}{2} \frac{q N_{drift}}{\varepsilon_0 \varepsilon_r} w^2_{scr,max} \tag{4.10}$$

Impact ionisation in semiconductor devices exhibits an exponential dependence on the electric field [11]. Using Fulop's approximation for the impact ionisation coefficients [12], the maximum width of the space charge region (at which avalanche breakdown occurs) can be analytically calculated to

$$W_{scr,bd} = \sqrt[8]{\frac{8}{1.8 \cdot 10^{-35}} \left(\frac{\varepsilon_0 \varepsilon_r}{q}\right)^7} N_{drift}^{-\frac{7}{8}}. \tag{4.11}$$

For silicon devices a set of analytical functions for the breakdown voltage, critical electric field and maximum width of the space charge region was derived assuming a constant doping concentration of the drift region [6]. For a specific dielectric constant of 11.8, the results are

$$\frac{E_{crit}}{\text{Vcm}^{-1}} = 4096 \cdot \left(\frac{N_{Drift}}{\text{cm}^{-3}}\right)^{-\frac{1}{8}}. \tag{4.12}$$

$$\frac{W_{scr,bd}}{\text{cm}} = 2.67 \cdot 10^{10} \cdot \left(\frac{N_{Drift}}{\text{cm}^{-3}}\right)^{-\frac{7}{8}}. \tag{4.13}$$

$$\frac{BV}{\text{V}} = 5.34 \cdot 10^{13} \cdot \left(\frac{N_{Drift}}{\text{cm}^{-3}}\right)^{-\frac{3}{4}}. \tag{4.14}$$

Note that the critical electric field slowly increases with increasing doping concentration. The well-known trade-off of the blocking voltage on one hand and the total device resistance on the other hand has led to the definition of a Figure-of-Merit for power semiconductor devices that is usually charted in a (double-logarithmic) $R_{DS,on}$ versus BV plot as shown in Fig. 4.9.

It is evident that, the integration density—being determined by the technology node—has a diminishing impact on the on-state resistance of LDMOS transistors. This is particularly true for devices with breakdown voltages beyond 50 V where the contribution of the channel towards the total device resistance becomes less prominent. Hence, More-Moore integration plays a minor role for these LDMOS transistors in smart-power ICs. This is primarily due to the requirement of a drift region significantly longer than the smallest feature sizes in state-of-the-art process technologies. The reasons are discussed in more detail in Chap. 5 of this textbook.

The one-dimensional silicon limit describes the correlation between the drift region resistance and the breakdown voltage for a pin-type structure. Since both properties depend on the doping concentration and the length of the drift region, an optimum for the drift region resistance at a given breakdown voltage can be derived. As will be shown, this optimum condition occurs in punch-through condition.

4.2.1.1 Drift Region Resistance under Non-punch Through Condition

A certain breakdown voltage can be achieved by several different combinations of doping concentrations and drift region lengths in punch-through condition. Consequently, the drift region resistance varies for different combinations as exemplarily depicted for a breakdown voltage of 50 V in Fig. 4.10.

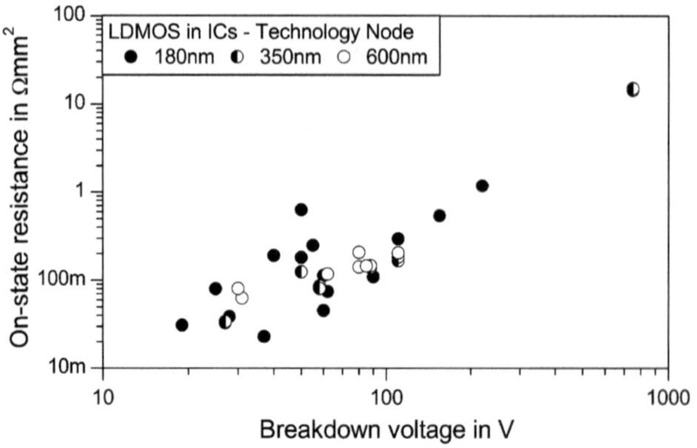

Fig. 4.9 $R_{DS,on}$ versus BV plot for different state-of-the-art LDMOS transistors in integrated circuits with different technology nodes

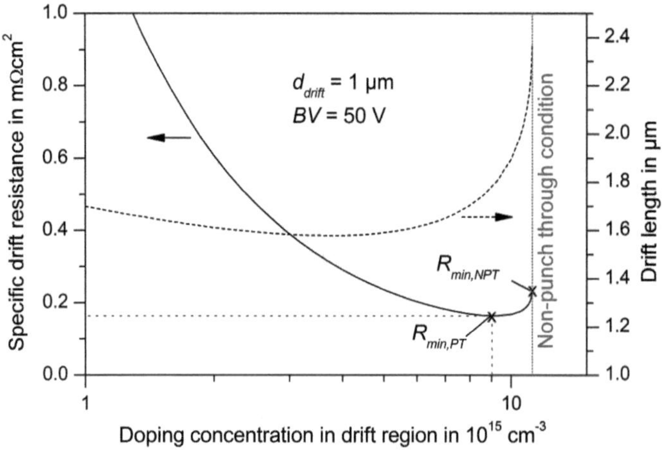

Fig. 4.10 Drift region resistance depending on doping concentration (and resulting drift region length in punch-through condition) for a given breakdown voltage of 50 V

In order to estimate the minimum drift region resistance of a pin-type drift region, the non-punch through condition is considered first, i.e. when the electric field at the end of the drift region (near the drain region) remains zero. From (4.14) a constant doping concentration of the drift region equal to

$$\frac{N_{drift,NPT}}{\mathrm{cm}^{-3}} = 1.98 \cdot 10^{18} \left(\frac{BV}{\mathrm{V}}\right)^{-\frac{4}{3}} \tag{4.15}$$

is required for the shortest possible drift region length in non-punch through condition. In the case of Fig. 4.10 this results in a doping concentration of

$1.12 \cdot 10^{16}$ cm^{-3}, and using (4.13) a drift region length (width of the space-charge region at breakdown) of 2.42 μm is necessary. If a non-punch through design is chosen, the trade-off between drift region resistance and breakdown voltage for silicon is given by

$$\frac{R_{drift,min,NPT} \cdot A}{\Omega \mathrm{cm}^2} = 1.55 \cdot 10^{-14} \cdot \left(\frac{\mathrm{cm}}{d_{Drift}}\right) \cdot \left(\frac{BV}{\mathrm{V}}\right)^{\frac{11}{3}}. \tag{4.16}$$

For the 50 V pin-structure in non-punch through condition a minimum drift region resistance of 0.24 mΩ cm^2 is calculated.

4.2.1.2 Drift Region Resistance in Punch-Through Condition

Another approach to reduce the drift region resistance is by shortening the length of the drift region (which also reduces the total length of the cell). Eventually, the space charge region will extend to the drain region and a punch-through condition of the electric field will occur as depicted in Fig. 4.11.

Now, the breakdown voltage that can be absorbed in the space charge region extending across the full length of the drift region L_{drift} is given by

$$BV = \frac{(E_{crit} + E_{drain})}{2} \cdot L_{drift}. \tag{4.17}$$

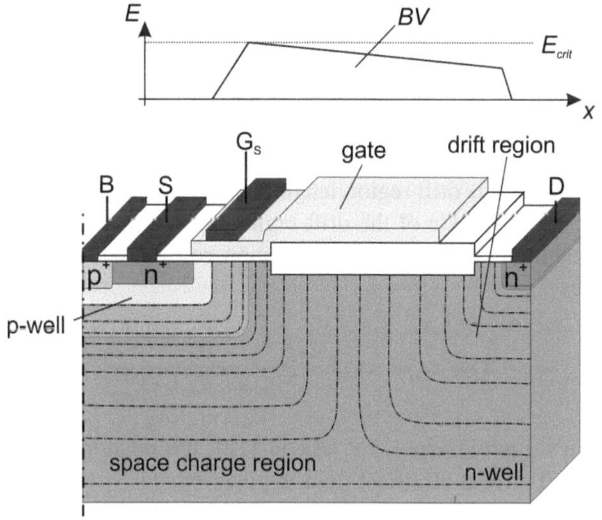

Fig. 4.11 LDMOS transistor with pin-type drift region in punch through condition—space charge region and electric field lines are simplified

Obviously, a further reduction of the drift region length will result in a lower breakdown voltage of the transistor. The electric field at the junction between drift region and drain contact can be expressed as

$$E_{drain} = E_{crit} - \frac{qN_{drift}}{\varepsilon_0 \varepsilon_r} L_{drift}, \tag{4.18}$$

yielding

$$BV = E_{crit} \cdot L_{drift} - \frac{qN_{drift}}{\varepsilon_0 \varepsilon_r} \frac{L_{drift}^2}{2}. \tag{4.19}$$

$$\leftrightarrow qN_{drift,PT} = \frac{2\left(E_{crit} L_{drift} - BV\right) \varepsilon_0 \varepsilon_r}{L_{drift}^2}. \tag{4.20}$$

Substituting this charge density along with the area of the device into (4.5) yields the minimum area-specific drift region concentration in a punch through design

$$R_{drift,min,PT} \cdot A = \frac{L_{Drift}^4}{2\left(E_{crit} \cdot L_{Drift} - BV\right) \mu_n \varepsilon_0 \varepsilon_r d_{drift}}. \tag{4.21}$$

For the pin-structure with a breakdown voltage of 50 V as shown in Fig. 4.10, a minimum drift region resistance of 0.164 mΩ cm^2 can be achieved using punch through condition. This constitutes to a reduction of over 30 % compared to the non-punch through design.

The optimum drift region length for a constant doping concentration is given by

$$L_{drift,min,PT} = \frac{4}{3} \frac{BV}{E_{crit}}, \tag{4.22}$$

which is 12.5 % longer than for a vertical power device with minimum drift region resistance [13]. With this drift region length, an electric field of 50 % of the critical field occurs at the drain-side of the drift region as shown in Fig. 4.12.

Then, the minimum area-specific drift region concentration in a punch through design becomes

$$R_{drift,min,PT} \cdot A = \frac{128}{27} \frac{BV^3}{\mu_n \varepsilon_0 \varepsilon_r d_{drift} \cdot E_{crit}^4}. \tag{4.23}$$

Using Fulop's approximation for silicon [12], the relationship between breakdown voltage and critical electrical field is given by

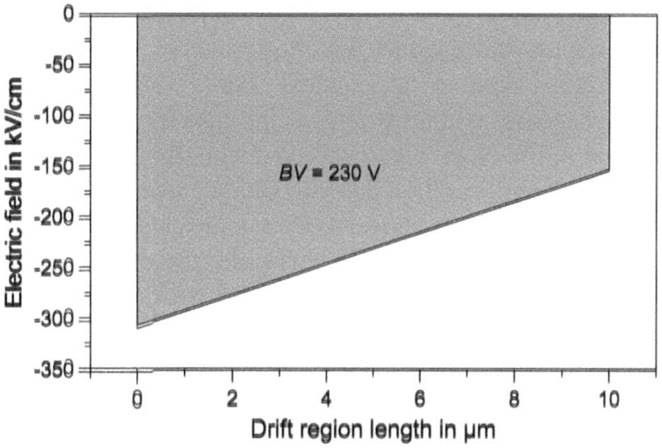

Fig. 4.12 Electric field distribution for a lateral pin-type drift region achieving the lowest theoretical drift region resistance for a drift region length of 10 µm

$$\frac{E_{crit,Si}}{\mathrm{Vcm}^{-1}} = 7.95 \cdot 10^5 \cdot \left(\frac{BV}{\mathrm{V}}\right)^{-\frac{1}{6}}, \tag{4.24}$$

and from (4.23) the trade-off between breakdown voltage and drift region resistance for silicon can be analytically expressed as

$$\frac{R_{drift,min,PT} \cdot A}{\Omega\mathrm{cm}^2} = 3.5 \cdot 10^{-15} \cdot \left(\frac{\mathrm{cm}}{d_{Drift}}\right) \cdot \left(\frac{BV}{\mathrm{V}}\right)^{\frac{11}{3}}. \tag{4.25}$$

These 1D silicon limits of the drift region resistance from (4.16) and (4.25) are compared to LDMOS transistors in commercially available smart-power ICs in Fig. 4.13.

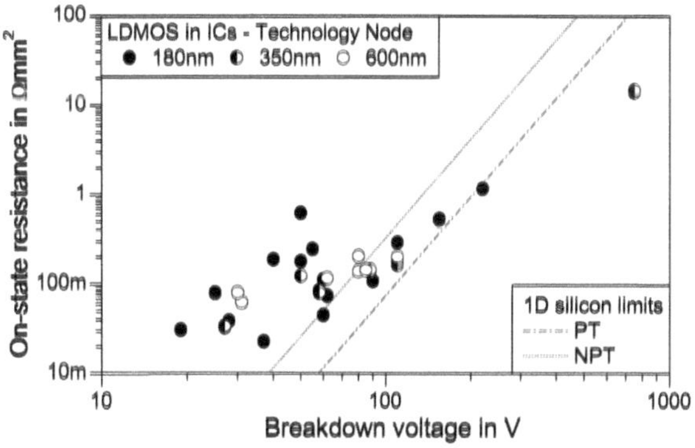

Fig. 4.13 Minimum drift region resistance versus breakdown voltage for punch through (PT) condition. The depth of the drift region is chosen to 1 µm. Typical drift region resistances of LDMOS transistors in commercially available Smart-Power ICs are included for comparison

As previously mentioned, the total device resistance can be approximated by the sum of the drift region and channel region resistances. There are two extremes which should be considered in this context. For a very short drift region, the channel resistance becomes dominant. This is the case for breakdown voltages well below 50 V as seen in Fig. 4.13. Obviously, reducing the drift region resistance for these devices will not significantly reduce the total device resistance. Conversely, a device with a breakdown voltage above 100 V requires a drift region which is much longer than the smallest feature size. In this case, the channel resistance can be neglected, and an improvement (reduction) of the drift region resistance will be beneficial towards the reduction of the total device resistance. It is also apparent from Fig. 4.13 that modern LDMOS transistors in integrated circuits are capable of breaking the 1D silicon limits. The RESURF design enabling such low drift region resistances is described in Chap. 5.

Regarding the minimization of static power losses in a circuit using LDMOS transistors, (4.25) in combination with (4.2) can be used to deduce the voltage dependence if the drift region resistance dominates:

$$P_{loss,on} = \frac{R_{DS,on}}{A} \cdot I_{on}^2 \cdot \delta = 3.5 \cdot 10^{-15} \cdot \left(\frac{\mathrm{cm}}{d}\right) \cdot \left(\frac{BV}{\mathrm{V}}\right)^{\frac{11}{3}} \cdot \frac{I_{on}^2}{A} \cdot \delta. \tag{4.26}$$

During the on-time t_{on} of each switching cycle, the current in the inductor of a switch-mode power supply rises to a peak value

$$I_{peak} = \frac{V_{in}}{L} \cdot t_{on}. \tag{4.27}$$

It is then discharged into the output capacitor during the remaining part of the operation cycle t_{period}. For the sake of simplicity, assuming discontinuous operation of a buck converter, the energy transferred through the inductor is given by

$$E_{trans} = \frac{1}{2} \cdot L \cdot I_{peak}^2 = \frac{1}{2} \cdot V_{in} \cdot I_{peak} \cdot t_{on}. \tag{4.28}$$

In order to transfer a certain amount of energy at a fixed duty cycle δ in a switch-mode circuitry also requires a fixed amount of power passing through the inductor. At a given switching frequency the amount of energy transferred by the inductance per switching cycle constitutes to

$$P_{trans} = \frac{1}{2} \cdot V_{in} \cdot I_{peak} \cdot \frac{t_{on}}{t_{period}} = \frac{1}{2} \cdot V_{in} \cdot I_{peak} \cdot \delta. \tag{4.29}$$

Since power semiconductor devices are operated with safety margin k_s of approx. 30–50 % below the breakdown voltage, the maximum operating voltage of an LDMOS transistor in a circuit at voltage V_{in} increases at the same rate as the breakdown voltage. Under this assumption, it can be inferred that

$$P_{trans} = \frac{k_s}{2} \cdot BV \cdot I_{peak} \cdot \delta. \tag{4.30}$$

Solving for the peak current, a relationship between the power losses and the breakdown voltage of an LDMOS transistor in a switch-mode power supply can be derived for a given transferable power supply as

$$P_{loss,on} \propto 3.5 \cdot 10^{-15} \cdot \left(\frac{\text{cm}}{d}\right) \cdot \left(\frac{BV}{\text{V}}\right)^{\frac{11}{3}} \cdot \frac{P_{trans}^2}{k^2 \cdot BV^2 \cdot A \cdot \delta^2} \propto BV^{\frac{5}{3}}. \tag{4.31}$$

Overall, the power losses incurred in a unipolar device with a uniformly doped (pin-type) drift region increase by a factor of five-thirds when increasing the device breakdown voltage a fixed amount of transferable energy. Larger device area must be utilized to counter this effect. In Chap. 5, these considerations will be used to show that it is favourable to increase breakdown voltage to reduce power losses for modern state-of-the-art LDMOS transistors.

Another approach for reduction of the total device resistance is the choice of a smaller technology node to reduce the channel length. Unfortunately, this method is only feasible for low blocking voltages where the resistance of the drift region is not dominant compared to the channel resistance. Moreover, care must be taken to prevent any reach-through condition: The space charge region does not only extend from the pn-junction into the drift region, but also into the p-well. The design of the p-well must implement a high enough doping concentration and a channel region wide enough to prevent the space charge region from reaching through to the source contact. Otherwise, and in contrast to the punch through condition, device breakdown will occur as electrons from the source are drifting into the space charge region, i.e. the blocking pn-junction becomes shorted.

It should be noted that power losses during the off-state of the switching cycle can be neglected as the leakage current through the transistor are several orders of magnitude lower than during the on-state compared to an increase in voltage by only one to two orders of magnitude.

A reduction in on-state resistance for a given device layout can be capitalized on by three different methods in switch mode power supplies which are depicted in Fig. 4.14.

4.2.2 Dynamic Losses: Device Capacitances and Switching Frequency

When operating frequencies are increased into the MHz range, charging and discharging device capacitances must be considered when evaluating the performance of LDMOS transistors. Three different capacitances of the transistor need to be

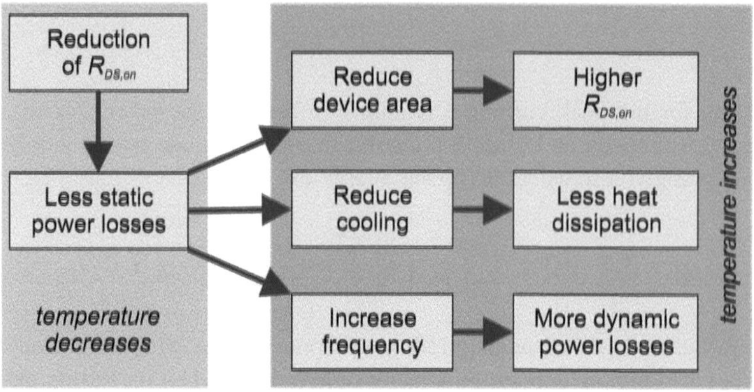

Fig. 4.14 Options for switch mode power supply designs based on reduction of on-state resistance in power transistors

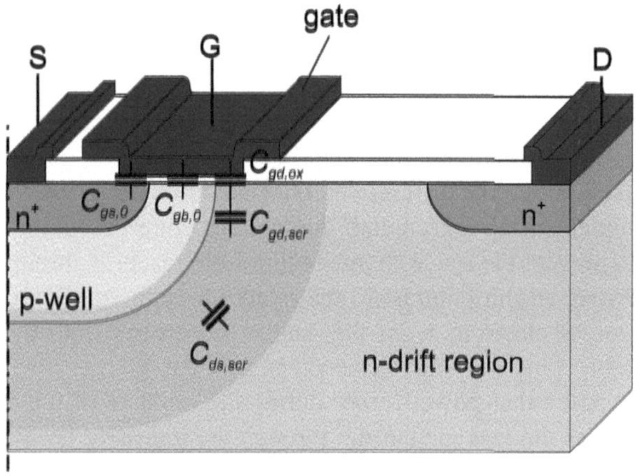

Fig. 4.15 Cross-section of LDMOS transistor with parasitic device capacitances

charged (turn-on) or discharged (turn-off) during each switching event. Using Fig. 4.15 these capacitances and their origin are introduced.

The gate-source capacitance C_{gs} arises from both the capacitance of the gate-source overlap $C_{gs,o}$ and the capacitance from the overlap of the gate over the body region $C_{gb,o}$. Note that source and body regions are shorted in most devices to suppress bipolar effect, e.g. turn-on of parasitic BJT. For a thin inter-metal dielectric, the input capacitance may increase further due to the coupling of the gate electrode and any passing source metallization (C_{imd}). In general, the gate-source capacitance is given by

$$C_{gs} = C_{gs,o} + C_{gb,o} = \frac{2x_p}{L_{cell}}\left(\frac{\varepsilon_0\varepsilon_r}{d_{gox}}\right). \tag{4.32}$$

The gate-drain capacitance C_{gd} occurs due to the overlap between the gate electrode and the drift region across the gate oxide.

$$C_{gd} = C_{gd,o} = \frac{2L_{gd,o}}{L_{cell}}\left(\frac{C_{gox}\cdot C_{gd,scr}}{C_{gox}+C_{gd,scr}}\right). \tag{4.33}$$

Here, it is assumed that the overlap for the polysilicon on the field oxide is relatively small compared to the contribution of $C_{gd,o}$, and this part of the gate-drain capacitance is neglected in designs without field plates. The impact of field plates on the device capacitance will be explained in Chap. 5. The area-specific capacitance of the field oxide is calculated by

$$C_{gox} = \frac{\varepsilon_0\varepsilon_r}{d_{gox}}. \tag{4.34}$$

Additionally, a space charge region below the field plate will reduce the capacitance due to the series connection with the space charge region in the semiconductor according to

$$C_{gd,scr} = \frac{\varepsilon_0\varepsilon_r}{w_{scr,vert}}. \tag{4.35}$$

The output capacitance emerges from the space-charge region between source/body and drain regions. The width of the space charge region depends on the drain-source potential, it can be approximated to

$$w_{scr,ds} = \sqrt{\frac{2\varepsilon_0\varepsilon_r(V_{DS}+V_{pn})}{qN_D}}. \tag{4.36}$$

Therefore, the drain-source capacitance equals

$$C_{ds} = C_{ds,scr} = \left(\frac{L_p}{L_{cell}}\right)\frac{\varepsilon_0\varepsilon_r}{w_{scr,ds}}. \tag{4.37}$$

An equivalent circuit for a typical LDMOS transistor including its parasitic device capacitances is shown in Fig. 4.16a).

For a comparison of transistors regarding switching operation, the total input capacitance C_{iss} with drain-source shorted, the reverse capacitance C_{rss} and the common source output capacitance C_{oss} with gate-source shorted are typically measured and published:

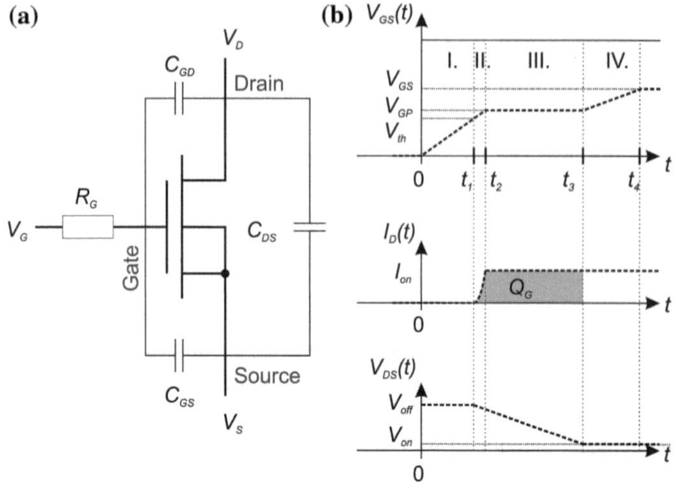

Fig. 4.16 **a** Equivalent circuit of LDMOS transistor showing the parasitic device capacitances; **b** current-voltage characteristics of LDMOS during device turn-on under clamped inductive switching

$$C_{iss} = C_{gs} + C_{gd} \tag{4.38}$$

$$C_{rss} = G_{gd} \tag{4.39}$$

$$C_{oss} = G_{gd} + C_{ds} \tag{4.40}$$

Switching the LDMOS transistor from off-state to on-state on an inductive load includes the following stages as illustrated in Fig. 4.16b):

First, the gate voltage is increased until the transistor starts to conduct (Phase I). During this process, both the gate-source capacitance C_{gs} and the transfer capacitance (at off-state) C_{gd} have to be charged (input capacitance). The speed of this charging process depends on the available gate current. The transistor starts to operate in saturation mode and the on-state drain current occurs at a gate voltage well below the maximum gate voltage (Phase II).

Then, the drain voltage is reduced towards the forward voltage drop in steady state (Phase III). Now, the gate current is primarily charging the gate-drain capacitance C_{gd} and the drain-source capacitance C_{ds} (output capacitance) while crossing the Miller-Plateau [14]. During this phase, the gate voltage only increases slightly to compensate the reduction of drain-source voltage in the saturation regime of the transistor.

As the transistor goes from the saturation region into the triode region, the gate voltage increases further to the maximum gate voltage by charging the gate-source capacitance and the transfer capacitance at on-state (Phase IV). At the end of this operation, the static total device resistance is reached.

4.2.2.1 Dynamic Losses from Switching in Switch Mode Converters

If the switching frequency in a switch mode power supply is high enough (typically in the 10 to 100 kHz range), the dynamic losses during switching that are primarily incurred due to the high voltages and currents at the Miller plateau become similar to the static losses during on-state of the transistor. The product

$$FOM = R_{DS,on} Q_{gd} \tag{4.41}$$

can be used as a Figure-of-Merit for different LDMOS transistors to judge the amount of dynamic power losses and the capability of the transistor to efficiently operate at high frequencies.

4.2.2.2 Dynamic Losses from Charging in Switch Mode Converters

Using a suitable design with low gate-drain capacitance, higher switching frequencies can be achieved. At switching frequencies close to the MHz range, charging and discharging of the input capacitance becomes another major source of power losses. The dynamic losses at a switching frequency f can then be described by

$$P_{SW} = C_{in} A_{cell} V_{GS}^2 f \tag{4.42}$$

The input capacitance is the sum of the gate-source and gate-drain capacitances. These dynamic power losses increase with device area. That is in contrast to the static power losses, where an increase in device area leads to a lower total device resistance. Therefore, for high switching frequencies, a trade-off between static and dynamic power losses exists. Small devices offer high switching speed with high static losses, while large devices have low static losses but high dynamic losses [6]. The total power losses also depend on the duty cycle δ and are given by

$$P_T = P_{on} + P_{SW} = \delta \cdot \frac{R_{DS,on}}{A} \cdot I_{on}^2 + C_{in} \cdot A \cdot V_{GS}^2 \cdot f. \tag{4.43}$$

It can be shown that the total power losses become minimal for an optimum area A_{opt} as follows:

$$A_{optimal} = \sqrt{\frac{R_{DS,on}}{C_{in}}} \left(\frac{I_{on}}{V_{GS}}\right) \sqrt{\frac{\delta}{f}}. \tag{4.44}$$

The minimum power losses are then calculated to

$$P_{minimal} = 2 I_{on} V_{GS} \sqrt{R_{DS,on} C_{in} \delta f}. \tag{4.45}$$

This trade-off is also illustrated in Fig. 4.17.

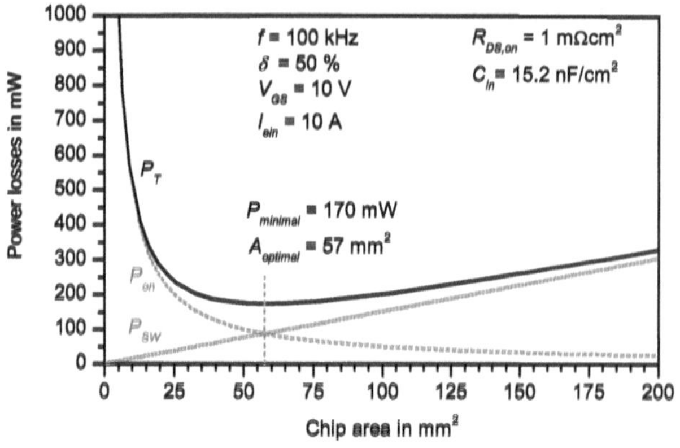

Fig. 4.17 Static and dynamic power losses in an LDMOS transistor depending on device area

Given that a decent LDMOS design for high frequency operation does not only require a low on-state resistance but also a low input capacitance, the correlations

$$FOM = \frac{R_{DS,on}}{C_{in}} \tag{4.46}$$

and

$$FOM = R_{DS,on} C_{in}. \tag{4.47}$$

are suitable Figures-of-Merit for comparison of different RF LDMOS transistor designs regarding area consumption and power losses respectively.

Remember that a higher switching frequency is also desirable from an application point of view in energy conversion because smaller reactance can be used in the periphery of the circuit.

4.2.3 Switching Frequency and Transistor Gain

In power amplifiers, the gate capacitance also has an impact on the current and power gains at high switching frequencies. The transition frequency of an LDMOS transistor which is a well-known value characterizing its high-frequency performance strongly depends on the gate capacitances. At the transition frequency the current gain of the transistor equals unity, i.e. it denotes the limit of current amplification.

At high switching frequencies, the current gain is dominated by charging (and discharging) the input capacitance. Using small signal analysis, the gate current at a high switching frequency f_{sw} in a MOS-gated transistor from a small change in gate voltage u_{gs} can be expressed as

$$i_g = 2\pi f_{sw} C_{in,app} \cdot u_{gs}. \tag{4.48}$$

Note that the apparent input capacitance of an amplifier is increased by the Miller effect [14]. The relationship

$$C_{in,app} = C_{gs} + C_{gd}(u_{gs}) \cdot (1 + R_{load} g_m) \tag{4.49}$$

can be derived for an LDMOS transistor with a resistive load R_{load} and the transconductance g_m. The gate-drain capacitance appears larger due to the feedback of the amplifier circuit on the input. The output current is related to the change in gate voltage by the transconductance.

$$i_d = g_m u_{gs}. \tag{4.50}$$

The current gain of the transistor is then found as

$$G_I = \frac{i_d}{i_g} = \frac{g_m}{2\pi f_{sw}\left(C_{gs} + C_{gd}(u_{gs}) \cdot (1 + R_{load} g_m)\right)}. \tag{4.51}$$

Increasing the switching frequency causes a linear increase in gate current. The transition frequency is denoted at which the gate current becomes equal to the drain current, i.e. the current gain equals unity (or 0 dB). Under this condition, the transition frequency is found by combining (4.48), (4.49), (4.50) and (4.51), yielding

$$f_T = G_I f_{sw}\big|_{G_I=1} = \frac{g_m}{2\pi\left(C_{gs} + C_{gd}(u_{gs}) \cdot (1 + R_{load} g_m)\right)}. \tag{4.52}$$

From this, it is obvious that a reduction of gate-drain capacitance is more beneficial than a reduction of gate-source capacitance with respect to high frequency performance. Taking it even further, it is feasible to accept more gate-source capacitance if gate-drain capacitance is reduced. The transition frequency can be extrapolated from a current gain versus frequency plot using the parameter h_{21} denoting the current gain into a short-circuited load.

From (4.52) it is also apparent, that the product of current gain and frequency $G_I f_{sw}$ is constant in this first-order approximation for a given device and circuit topology. It directly translates to the requirements of the gain-bandwidth-product as defined in Chap. 3, highlighting the relevance of the transition frequency as a performance indicator.

In order to also account for the impact of device resistances, the maximum oscillation frequency f_{max} can be examined. It is defined as the frequency at which the maximum available gain becomes unity (or 0 dB) which is a fundamental requirement for oscillation. The maximum oscillation frequency depends on the gate-charging, source, gate and output resistances R_i, R_s, R_g, R_{ds}, respectively [15]. It can be calculated according to:

$$f_{max} \approx \frac{f_T}{2\sqrt{\frac{R_i+R_s+R_g}{R_{ds}} + 2\pi f_T R_g C_{gd}}}. \tag{4.53}$$

However, extraction of the maximum oscillation frequency is more difficult and requires measurement of maximum stable and maximum available gain. A more detailed discussion on maximum oscillation frequency is given by Davis [16].

4.2.4 Switching Frequency and Output Power

Another quantity that determines device performance for operation at radio frequencies is the source impedance. The necessity of a low source inductance L_S of an RF LDMOS transistor can be derived from the analysis of its small-signal equivalent circuit model shown in Fig. 4.18.

The power gain as defined by (3.1) is found for this transistor to be

$$G_P = \frac{R_{load} g_m^2}{\left[(C_{GS} + C_{GD}(1 + R_{load} g_m))^2 + g_m L_S(C_{GS} + C_{GD} + C_{GD} R_{load} g_m)\right](2\pi f)^2}, \tag{4.54}$$

at operation frequency f [17]. Hence, a low inductive coupling of the source electrode to the circuit board (minimizing L_s) yields a higher power gain. The drain inductance is usually considered to be small compared to the inductance of the external load so it is neglected. Nevertheless, drain inductance can be considered, e.g. to construct load-pull power contours [18]. From (4.54) it is also apparent that the power gain reduces with the square of the operating frequency. An increase

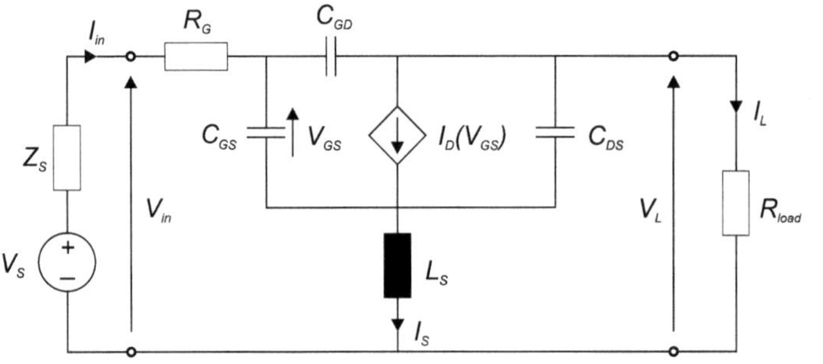

Fig. 4.18 Small-signal equivalent circuit model of LDMOS transistor, adopted from [17]

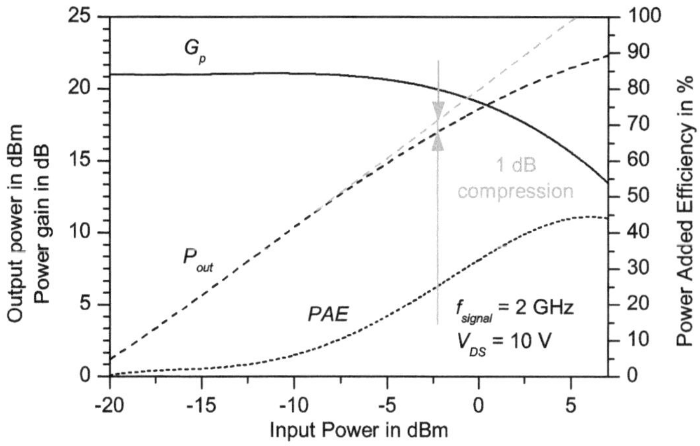

Fig. 4.19 Measurement of output power of an integrated RF LDMOS transistor, data from [19]

of the frequency by a factor of 10 results in a reduction of power gain by a factor of 100 or by −20 dB. The power gain at a given output frequency can be evaluated by measurement of output power over a wide range of signals with different input power as depicted in Fig. 4.19.

Considering the power provided by the biasing network, the power added efficiency is calculated. Power gain and PAE are typically examined at the 1 dB compression point (see Fig. 3.7).

Silicon-based RF transistors are capable of operating in the GHz range with output power up to some 100 W. In order to achieve higher operating frequencies or higher output power, a vast number of different solid-state and vacuum devices have been implemented over the last century. Figure 4.20 summarizes these developments.

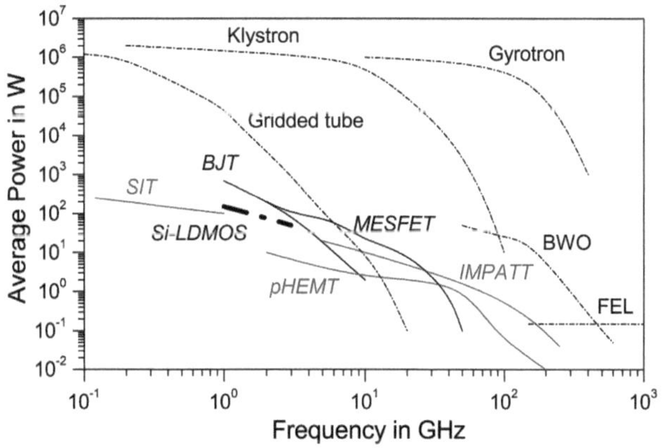

Fig. 4.20 Average RF output power versus frequency for various semiconductor and vacuum devices, adopted from [20] and [21]

In order to minimize the matching effort at the output of the power amplifier, its output resistance should be close to 50 Ω. For high output power, increasing operating voltage is required according to (3.3). The breakdown voltage of a transistor is closely related to its maximum operating voltage. The capability of an RF amplifier to provide high output power and gain at high operating frequency can be estimated from the ratio of breakdown voltage and maximum oscillation frequency.

4.2.5 Power Densities and Long Term Stability

Even though the volume of a semiconductor chip is not significantly contributing to the volume or weight of a power electronic system, the chip area is of high importance regarding system expenditure. This is due to the fabrication costs of such a sophisticated high-end product. It is therefore desirable to use the smallest possible chip area for a given product. What constitutes to the smallest size of a smart-power IC or power semiconductor device is determined by both power densities and reliability: For a given device design, packaging and circuit topology, a semiconductor device will generate power losses. This heat has to be dissipated through the package or module assembly. According to the associated thermal resistance, the package is capable of dissipating a certain power density. Hence, a certain chip area is required to prevent the chip from incurring excessive junction temperatures. The junction temperature in turn is limited by the long term stability requirements of the application and thus the device. While there are exceptions, e.g. HVDC transmission, the lifetime of a semiconductor device is expected to match the specified operating time of the system. This is particularly true, where a replacement of failed device is not possible as in space applications, aviation, power conversion or feasible like with consumer electronics.

4.2.6 Design and Development of LDMOS Transistors in Smart Power ICs

The design of LDMOS transistors usually starts with TCAD-based device simulations in a given process flow. The process flow can either be based on a standard CMOS process with the goal of developing a high-voltage extension to this technology or on an existing device design with the goal of further improvement, e.g. by implementation of optimized technologies introduced in Chap. 5 through Chap. 8. Hereby, the feasibility of the fundamental device geometry and doping concentrations is verified. These simulations can also serve as a starting point for device modelling (e.g. in SPICE). Moreover, the suitability of the junction termination can be investigated.

Next, the photo mask layout is extracted from the evaluated geometry. The layout can include power transistors, the logic (analog/digital) part and memory cells. It will also contain contact pads and protective devices against electrostatic discharge. The extent of the design may vary from only a small number of power devices (in the case of an integrated half-bridge) up to full-fledged microcontrollers which control several power stages.

Then, the device design is fabricated for a limited number of devices using an engineering run. Electrical characterization of static and dynamic device properties is conducted to judge the performance of the fabricated devices. This judgment can be augmented by a comparison to the TCAD simulation results.

Based on the results and possibly a failure analysis, a redesign is required based on the steps previously discussed. Once adequate device performance is achieved, reliability tests are typically conducted to explore the ruggedness and boundary conditions for operation.

4.3 Fundamental Device Topologies of Lateral Power Semiconductor Devices

Based on the above mentioned trade-offs two main device topologies for LDMOS transistors have evolved from the initial design from Sigg et al. [3].

- The *lateral power MOSFET* is implemented in smart-power ICs for energy conversion applications.
- The *lateral RF MOSFET* operates in RF amplifier circuits at high switching frequencies for wireless information transmission.

Similarities and differences between these two designs will be discussed next.

4.3.1 Lateral Power MOSFETs in Smart-Power ICs

Smart-power ICs enable applications which require power conversion. The power devices implemented in these circuits have to exhibit low power losses. In contrast to RF applications, high switching frequency is less important. In Fig. 4.21 a state-of-the-art LDMOS device for this kind of application is depicted where the drain-source voltage is blocked laterally.

In this device, the drain, gate and source contacts are all situated on the same semiconductor surface. Moreover, the backside of the wafer can be electrically isolated making it a decent choice for integration into smart-power ICs where the source electrode is not biased to ground potential. Also, the source and drain metallization of the device have to be intermitted. This reduces the area of metallization compared to a vertical device because design rules dictate that certain

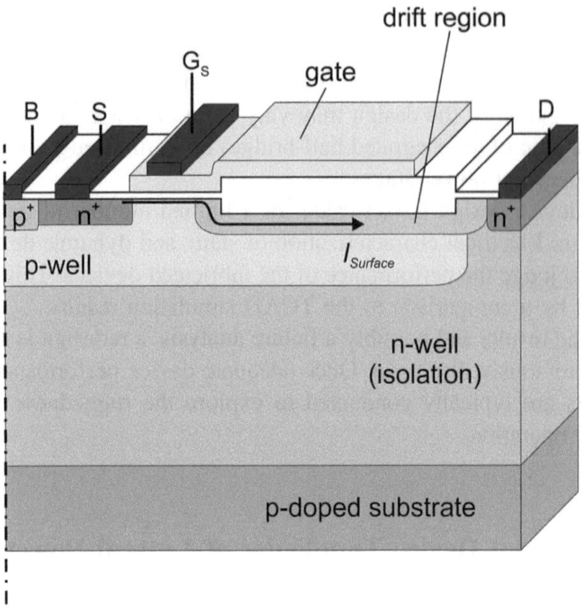

Fig. 4.21 Cross section of an LDMOS transistor for integration into smart-power ICs

distances between the metallization lines have to be obeyed. In turn, the current densities are considerably higher than in the VDMOS design and the effect of electromigration needs to be considered more carefully.

Using advanced techniques like RESURF design and field plates and a reduction in feature size, the total device resistance of these LDMOS devices has been significantly reduced in the past decade. These techniques employed in state-of-the-art LDMOS transistors for smart-power ICs will be explained in Chap. 5.

4.3.2 LDMOS Transistors Optimized for Radio-Frequency Applications

For the operation under switching frequencies in the GHz range, a different LDMOS topology is used. This device is optimized towards small device capacitances. Especially a small transfer capacitance between gate and drain is crucial for achieving high operating frequencies. Therefore, a ground shield between gate and drift region is routinely implemented. The cross-section of an RF LDMOS is shown in Fig. 4.22.

Besides, the source contact can be transferred from the top to the bottom of the transistor by a p^+-sinker to reduce the source inductance. These techniques are described in Chap. 5 in more detail.

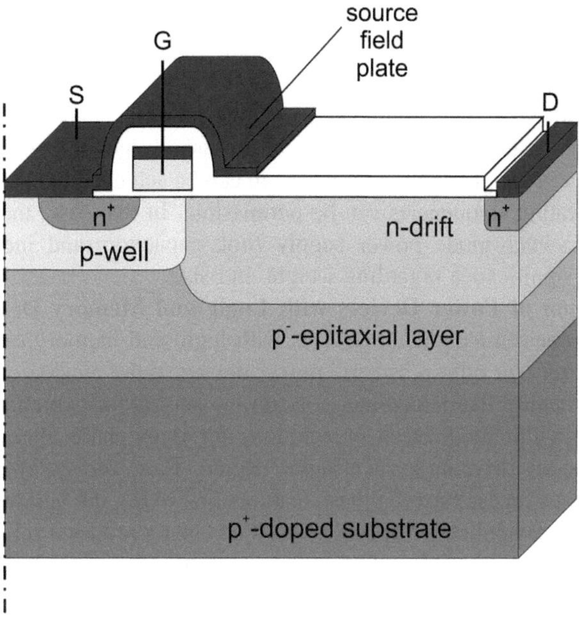

Fig. 4.22 Cross section of an LDMOS for integration in RF CMOS technology

4.4 Aspects of Integration of Power Transistors in ICs

The integration of lateral power transistors into integrated circuits, e.g. as application-specific integrated circuits, yields several benefits and disadvantages that are discussed next.

4.4.1 Advantages of LDMOS Transistors in Integrated Circuits

Integrating lateral power transistors in integrated circuits offers several distinct advantages over a discrete assembly, e.g. on a printed copper board.

Reliability and Robustness With respect to electrical connections like wires or metal routes on a printed copper board, the metallization lines in a semiconductor chip show lower failure rates. Additionally, less solder and bonding joints are required when several building blocks are integrated into a System-on-a-Chip (SoC). Therefore, reliability of integrated circuits is superior to that of a circuit composed of discrete building blocks.

Similarly, the robustness of a circuit built on a printed copper board to overvoltage and current surges is low due to parasitic inductances and capacitances from the leads, wires and packages. Integrated circuits better withstand extreme electrical conditions while also being less prone to failure due to high ambient temperatures, vibration and shock.

Weight and Size While integrated circuits are small and light-weight, these benefits compared to a discrete component are usually negligible. Nevertheless, a large reduction in weight and size of a system can be achieved by integrated circuits as higher operating frequencies can be permissible. In that case, the passive components in a switch-mode power supply (link capacitors and inductor) can be reduced with significance regarding weight and size.

Combination of Power Devices with Logic and Memory Devices By combining the power semiconductor devices with logic and memory circuits on chip, the manufacturer can offer a "smart" power device to the outside world. This can include functionality like temperature or current sensors for overcurrent shutdown, on-chip pulse-width modulation or solutions for three-phase alternating currents which can directly drive an asynchronous engine. Then, the system designer does not have to focus on establishing these features first. Also, the building blocks used in the design of an application specific circuit can be reused for a subsequent design or a new product release. In addition, the design components of an application specific integrated circuit are not as easily identified as on a printed copper board. This includes both the functional building blocks and the interconnection of these blocks to form the circuit.

4.4.2 Limitations and Drawbacks of LDMOS Transistors in Integrated Circuits

With the integration of unipolar power devices in integrated circuits a number of drawbacks are associated.

Consequences of Unipolar Charge Transport in Lateral Devices Due to the unipolar charge transport, conductivity modulation like in bipolar transistors cannot be exploited in LDMOS transistors. As vertical power transistors cannot readily be used in smart-power ICs, the device current has to be passed through the source and drain metallization on the same side of the semiconductor. This results in the need for smaller, typically finger-shaped, metallization contacts requiring a larger footprint of the cell. In summary, LDMOS transistors incur higher static power losses than stand-alone VDMOS and UMOS transistors. Also, channel and drift region resistance increases with temperature in unipolar devices due to the associated reduction of channel and bulk mobility.

Field Enhancement at the Device Surface When the drain-source voltage is blocked in a planar transistor with a lateral drift region, additional effort (e.g. a gate field plate) is necessary to prevent premature avalanche breakdown and oxide degradation at the device surface. This design requirement results in increased coupling capacitance between gate and drain of the transistor and hampers operation at radio frequencies. For this reason, LDMOS transistors for energy conversion and for RF applications have taken different design routes.

Combination of Power Devices with Logic and Memory Devices Smart-power ICs are often tailored towards a particular application. Therefore, system designers can be faced with a potential solution that does not fit their needs because some undesired functionality has also been hardwired into the IC independent of the choice of the power semiconductor devices. Conversely, the available options in such an IC may not be sufficient for particularly demanding applications. If the circuit is not properly designed, it cannot even be reused, e.g. when core functionalities in external libraries cannot be amended to the new task at hand.

4.5 Safe Operating Area for Lateral Power Transistors

The maximum currents and voltages that can reliably be applied to lateral power transistors are limited due to several physical effects. These physical effects are well understood and have already been described in detail in various textbooks before. Therefore, a review is omitted in this textbook. Instead, any effects that have not been introduced in the previous sections are referenced by suitable references dealing with these effects. Operating conditions of lateral power transistors are primarily limited by the following conditions and effects:

A maximum drain-source voltage may not be exceeded to prevent avalanche breakdown due to impact ionisation. This limitation has been described in this chapter in great detail.

The static power losses must be limited allowing for continuous dissipation of the generated heat without thermally overheating the transistor. Drain current reduces with increasing temperature in a unipolar device [22]. Hence, the thermal runaway phenomenon known from bipolar transistors due to an increase in gain over temperature does not occur [23]. Nevertheless, the junction temperature $T_{Junction}$ in relation to the ambient temperature $T_{ambient}$ must be kept in check to prevent melting of the solder or the bond wires which would disconnect the silicon devices from its surrounding.

$$P_{loss} = \frac{T_{Junction} - T_{ambient}}{R_{Th,DC}}. \tag{4.55}$$

The static power losses in the semiconductor can be increased when the static thermal resistance $R_{Th,DC}$ is reduced. Under pulsed operation, which is typically occurring in switch-mode circuits, less power losses are incurred. For an accurate calculation, the dynamic thermal resistance has to be considered. Often, the dynamic thermal coefficient R_K of the chip package is listed in datasheets. The safe operating area of a typical LDMOS transistor in a chip package is shown in Fig. 4.23.

The shaded region (high currents at low drain-source voltages) represents an inaccessible region given by the on-state resistance of the LDMOS transistor.

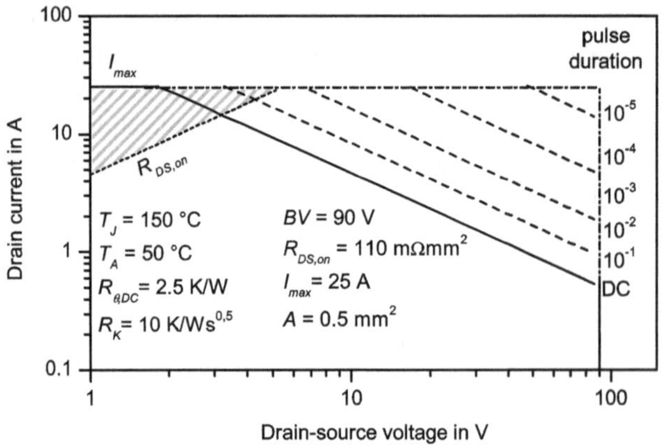

Fig. 4.23 Safe operating area of a packaged LDMOS transistor

The maximum rated device current is limited by the current handling capability of the bond wires.

Moreover, the maximum gate voltage is limited by time-dependant dielectric breakdown [24]. Activation energies of dielectric breakdown can be determined from constant current and constant voltage stress (depending on the thickness of the dielectric). In order to ensure sufficient lifetime of the gate dielectric, mission profiles based on these activation energies are routinely employed.

For power devices in general and integrated LDMOS transistors in particular, hardness to electro-migration must be considered. In lateral power transistors, the drain and source contacts are located at the same side of the semiconductor, effectively doubling the current density compared to a vertical power device. Often, multiple metal layers in an integrated process flow are paralleled to cope with the high current densities passing through the source and drain metallization layers [25].

Finally, immunity to hot carrier injection (HCI) should be mentioned. Due to the high electric fields in the drift region adjacent to the drain side of the gate electrode, an unfavourable design may give rise to the injection of hot carriers into the gate dielectrics. In the case of an n-channel LDMOS transistor, these electrons can cause an increase in threshold voltage and even damage the gate dielectric over time resulting in premature breakdown [26, 27].

4.6 More-Moore and More-than-Moore Integration Methodology in Smart-Power ICs and RF Amplifier Technologies

Due to the significant impact of the drift region length on breakdown voltage, smart power ICs for high operating voltages barely benefit from a reduction of smallest feature size (More-More integration). Here, different concepts like the implementation of field plates and the RESURF principle have already driven performance improvements over the last decades. Smart-power ICs operating above 20–30 V clearly benefit from the shrinkage of power semiconductor devices associated from these developments following More-than-Moore integration.

At the same time, smart-power ICs are also affected by the reduction of smallest feature size (More-Moore approach) because the logic circuitry becomes smaller. Then, however, the area demand of the power semiconductor devices in relation to the logic part increases and more area per wafer has to be sacrificed for "bulky power devices" using an expensive high resolution technology. This increases the demand for highly integrated LDMOS transistors with every new technology node.

For RF power amplifiers and low voltage smart-power ICs, the situation is somewhat different. Minimization of parasitic capacitances is required, and high resolution lithography is a convenient way to achieve this. In relation to the logic building blocks, LDMOS transistors for low operating voltages still benefit from the device shrinkage to some extent. In this case, the call for an implementation of More-than-Moore approaches for further device shrinkage is not so imminent. However, a reduction in gate length gives rise to short channel effects that can deteriorate the performance of RF transistors. In order to prevent performance degradation, additional process steps like lightly doped drain implantation have been implemented in CMOS technology and carried over to the development of RF LDMOS transistors [28]. It should also be noted, that there is also a theoretical limit to device current, independent of channel length and mobility due to channel back-scattering in highly integrated transistors [29].

References

1. G.E. Moore, Cramming more components onto integrated circuits. Electron. Mag. **38**(8), 114–117, (1965)
2. International Technology Roadmap on Semiconductors, 2013. http://www.itrs.net. (Zugriff am 31 05 2014)
3. H.J. Sigg, G.D. Vendelin, T.P. Cauge, J. Kocsis, D-MOS transistor for microwave applications. IEEE Trans. Electron Devices **ED-9**(1), 45–53 (1972)
4. G. Deboy, M. März, J.-P. Stengl, H. Strack, J. Tihanyi, H. Weber, New generation of high voltage MOSFETs breaks the limit line of silicon, in *International Electron Device Meeting*, San Francisco, USA (1998)
5. S.M. Sze, K.K. Ng, *Physics of Semiconductor Devices* (John Wiley, Hoboken, 2007)
6. B.J. Baliga, *Fundamentals of Power Semiconductor Devices* (Springer, New York, 2008)
7. E.H. Nicollian, J.R. Brews, *MOS* (*Metal Oxide Semiconductor Physics and Technology*) (Wiley, Hoboken, 2002)
8. S. Takagi, A. Toriumi, M. Iwase, H. Tango, On the universality of inversion layer mobiliy in Si MOSFET's: part i-effects of substrate impurity concentration. IEEE Trans. Electron Devices **41**(12), 2357–2362 (1994)
9. K.G. McKay, Avalanche breakdown in silicon. Phys. Rev. **94**(4), 877–884 (1954)
10. J.L. Moll, R. van Overstraeten, Charge multiplication in silicon pn junctions. Solid-State Electron. **6**, 147–157 (1963)

11. S.M. Sze, G. Gibbons, Avalanche Breakdown voltages of abrupt and linearly graded pn junctions in Ge, Si, GaAs and GaP. Appl. Phys. Lett. **8**(5), 111–113 (1966)
12. W. Fulop, Calculation of avalanche breakdwon of silicon p-n junctions. Solid-State Electron. **10**, 39–43 (1967)
13. C. Hu, Optimum doping profile for minimum ohmic resistance and high-breakdown voltage. IEEE Trans. Electron Devices **ED-26**(3), 243–244 (1979)
14. J.M. Miller, Dependence of the input impedance of a three-electrode vacuum tube upon the load in the plate circuit. Sci. Pap. Bur. Stan. **15**(351), 367–385 (1920)
15. J.W. Chung, W.E. Hoke, E.M. Chumbes, T. Palacios, AlGaN/GaN HEMT with 300-HGz fmax. IEEE Electron Device Lett. **31**(3), 195–197 (2010)
16. R. Davis, GaAs FETs—physics, design, and models, in *Handbook of RF and Microwave Power Amplifiers* (Cambridge University Press, Cambridge 2012), pp. 42–102
17. M.M. De Souza, P. Fioravanti, G. Cao, D. Hinchley, Design for reliability: the RF power LDMOSFET. IEEE Trans. Device Mater. Reliab. **7**(1), 162–174 (2007)
18. S.C. Cripps, A theory for the prediction of GaAs FET load-pull power contours, in *International Microwave Symposium* (Boston, USA, 1983)
19. R. Sorge, A. Fischer, A. Mai, P. Schley, J. Schmidt, C. Wipf, T. Mausolf, R. Pliquett, R. Barth, K.E. Ehwald, Complementary RF LDMOS module for 12 V DC/DC converter and 6 GHz power applications, in *Silicon Monolithic Integrated Circuits in RF Systems* (Phoenix, USA, 2011)
20. V.L. Granatstein, R.K. Parker, C.M. Armstrong, Vacuum electronics at the dawn of the twenty-first century. IEEE Proc. **87**(5), 702–716 (1999)
21. R.J. Trew, SiC and GaN transistors—is there one winner for microwave power applications? IEEE Proc. **90**(6), 1032–1047 (2002)
22. G.M. Dolny, G.E. Nostrand, K.E. Hill, The effect of temperature on lateral DMOS transistors in a power IC technology. IEEE Trans. Electron Devices **79**(4), 990–995 (1992)
23. J.D. Cressler, D.D. Tang, K.A. Jenkins, G.-P. Li, E.S. Yang, On the low-temperature static and dynamic properties of high-performance silicon bipolar transistors. IEEE Trans. Electron Devices **36**(8), 1489–1502 (1989)
24. D.J. Dumin, *Oxide Reliability: A Summary of Silicon Oxide Wearout, Breakdown, and Reliability* (World Scientific, Singapore, 2002)
25. D.C. Burdeaux, W.R. Burger, Intrinsic reliability of RF power LDMOS FETs, in *International Reliability Physics Symposium*, (Monterey, USA, 2011)
26. A. Acovic, G. La Rosa, Y.-C. Sun, A review of hit-carrier degradation mechanisms in MOSFETs. Microelectron. Reliab. **36**(7–8), 845–869 (1996)
27. A.W. Ludikhuize, M. Slotboom, A. Nezar, N. Nowlin, R. Brock, Analysis of hot-carrier-induced degradation and snapback in submicron 50 V lateral MOS transistors, in *International Symposium on Power Semiconductor Devices & ICs* (Weimar, Germany, 1997)
28. O. Bengtsson, A. Litwin, J. Olsson, Small-signal and power evaluation of novel BiCMOS-compatible short-channel LDMOS technology. Trans. Microw. Theory Tech. **51**(3), 1052–1056 (2003)
29. F. Assad, Z. Ren, S. Datta, M. Lundstrom, P. Bendix, Performance limits of silicon MOSFET's, in *International Electron Devices Meeting* (Washington, USA, 1999)

Chapter 5
Modern MOS-Based Power Device Technologies in Integrated Circuits

The state-of-the-art of semiconductor device technology for LDMOS transistors in integrated circuits is introduced in this chapter. First, a review on the history of lateral power transistor development is presented. The evolution of the lateral power MOSFET is depicted and compared to the bipolar junction transistor. The operation principle of lateral power MOSFETs is reviewed including forward, reverse and switching operation. Then, different design concepts that have enabled the breakthrough for lateral power MOSFETs are presented: A key element of modern lateral power MOSFETs is the implementation of the *reduced surface field* concept in addition to field plates allowing for a decent trade-off between operation voltage and power losses. These concepts can be readily implemented in CMOS-based integrated circuits with little additional fabrication effort. Nevertheless, the design has to be carefully optimized to yield best-in-class performance. In particular, multi-acting RESURF designs appeal due to lower device losses but fabrication is more difficult. And the field plate concept offers an additional degree of freedom in the design. Depending on the circuit topology in which the power transistor is used, high-side switching capability can be designed in. Finally, stability against avalanche breakdown can be achieved by suitable device design. Similar considerations also hold true for lateral RF MOSFETs in integrated circuits, where RESURF designs are combined with ground shields and source sinkers to minimize device resistance and capacitance. Here, immunity to hot carrier injection and electromigration effects need to be considered as well. The Figures-of-Merit defined in Chap. 4 will be reviewed with respect to their applicability to RESURF-based LDMOS transistors. A comparison of different device technologies establishes an overview regarding the development of LDMOS transistors in integrated circuits.

5.1 History of Lateral Power Transistor Development

Concerning power semiconductor devices there are two distinct traits based on the carriers responsible for charge transport that have to be contemplated: On one hand bipolar devices like bipolar junction transistors and thyristors emerged in the 1950s,

T. Erlbacher, *Lateral Power Transistors in Integrated Circuits*,
Power Systems, DOI 10.1007/978-3-319-00500-3_5

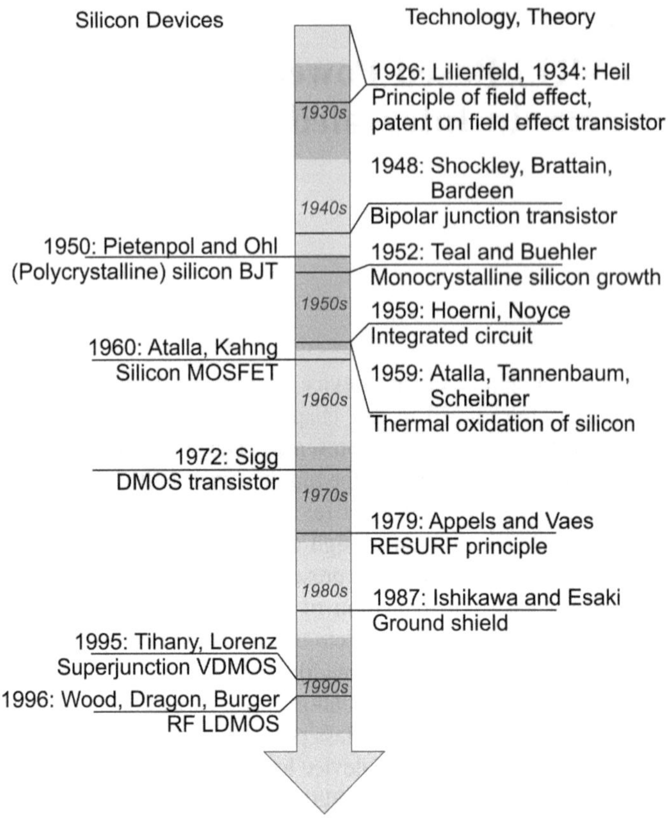

Fig. 5.1 Timeline of important developments towards bipolar and unipolar lateral power transistors

and on the other hand unipolar devices like lateral and vertical DMOS were introduced in the 1970s. A timeline related to the milestones in development of lateral power transistors is given in Fig. 5.1.

Chronologically, the *bipolar junction transistors* spearheaded the development of semiconductor industry offering a prospective replacement for vacuum tubes. In 1948, Brattain and Bardeen demonstrated the first solid state amplifier, a point-contact bipolar transistor on germanium [1] and Shockley provided the theory behind it [2]. The first silicon transistor on polycrystalline silicon was demonstrated by Pietenpol and Ohl in 1950 [3]. However, amplification was still poor due to the high defect density of the available semiconductor material. This started to change when in 1952 Teal and Buehler grew the first monocrystalline silicon substrate which exhibited significantly lower defect densities than polysilicon material [4]. So finally, silicon transistors based on monocrystalline silicon came into production around 1954. The success of silicon technology was also related to the manufacturability of stable silicon dioxide layers with low defect densities.

In 1959, the work of Hoerni, realizing the first transistor in planar technology, paved the way for integrated circuit technology as we know it today [5]. Noyce built his *monolithic integrated circuit* concept on Hoerni's planar technology which he filed as a patent on "Semiconductor device-and-lead structure" the same year [6].

By the inclusion of a lightly doped collector region, bipolar junction transistors with high operating voltages could be manufactured. In this topology, charge modulation of the BJT in on-state occurs, effectively flooding the lightly doped region with a large amount of electrons and holes. This allows for high current densities at low power losses.

However, the bipolar junction transistor suffers from several drawbacks for power electronic applications: First of all, charge modulation limits the switching frequency of BJTs for power electronic applications because during device turn-off the excess carriers need to be removed (by diffusion or recombination) before a space-charge region can extend in the device to block the collector–emitter voltage. Additionally, the charge modulation has to build up when turning on the device by diffusion of minority charge carriers in the drift region. Both effects result in significant dynamic switching losses compared to power devices that rely on unipolar, i.e. majority carrier based, charge transport.

Next, the BJT is a current controlled device necessitating a rather complex drive circuit for the base with high continuous drive power compared to MOS gated devices. This demand results in significant input power losses incurring in amplifying circuits.

Also, the BJT exhibits a negative temperature coefficient of the electrical device resistance. When the junction temperature of a BJT increases, it will start conducting more current. This in turn, increases the heat dissipation, further increasing the temperature of the device. If this condition is not well controlled by adequate cooling, the device gets destroyed by thermal runaway. This effect hampers parallelization of multiple BJTs to increase the output power of a power electronic circuit.

After bipolar transistors were successfully introduced by semiconductor industry, the *unipolar transistor* launched another wave of innovation for semiconductor based solutions. The operation principle of a transistor based on the field effect was discovered by Lilienfeld in 1926 [7], well before theory on BJTs was released. Based on that, the function of a field effect transistor was described by Heil in 1934 [8]. But at that time, fabrication of such field effect transistors failed due to insufficient pureness of semiconductor materials. For operation of a surface channel based device, the interface state density of the semiconductor–oxide interface could not be properly controlled. Additionally, the gate dielectric did not exhibit sufficient strength against electric fields due to defects and impurities.

This started to change with advances in silicon growth as previously described on one hand, and with the thermal oxidation of silicon introduced by Atalla et al. in 1959 on the other hand [9]. So in 1960, Atalla and Kahng succeeded in manufacturing the first operating silicon MOSFET [10]. But again, due to the lack of control over impurities, it took another 15 years until MOSFETs could be routinely used for electronics [11].

For power semiconductor devices, further milestones have to be considered. The first work on *DMOS* transistors for microwave applications was published by Sigg in 1972 [12]. And in 1979, the performance of these DMOS transistors towards power electronic applications were enhanced by the discovery and understanding of the *RESURF*-principle by Appels and Vaes using thin epitaxial layers [13]. They discovered that shallow pn-junctions can exhibit higher lateral breakdown voltages than deeper junctions with otherwise similar properties. The effects of RESURF-based drift region parameters on the static properties of LDMOS transistors were then presented by Colak just 2 years later [14]. The RESURF design uses charge balancing to locally increase the effective doping concentration, lowering the drift region resistance without diminishing the breakdown voltage. This technology was also amended in a junction termination technique called *junction termination extension* by Temple in 1977 [15].

The *ground shield* that leads to a tremendous reduction of transfer capacitance in LDMOS transistors was first presented by Ishikawa and Esaki in 1987 for VDMOS transistors operating at 900 MHz [16]. This development was carried over to manufacture *RF LDMOS* transistors for operation beyond 1 GHz by Wood et al. almost 1 decade later [17].

Later on, another leap in device performance was achieved by the introduction of the *superjunction* MOSFET by Tihany and Lorenz in 1995 [18] for vertical power devices. Basically, it resembles a segmented RESURF design with alternating p-doped and n-doped "pillar" regions. The concept will be discussed in Chap. 6 with respect to its application in lateral power transistors for its capability of further reducing drift region resistance.

5.2 LDMOS Transistors for Smart Power ICs

Compared to bipolar transistors, unipolar MOS-based power semiconductor devices offer the advantage of fast switching and a voltage controlled gate, i.e. the gate drive circuit is only required to provide enough current to charge and discharge the device capacitances during switching. Moreover, owing to the positive temperature coefficient of the drift region resistance, parallelization of unipolar power semiconductor devices is possible. A local increase in device temperature results in reduced local currents, which in turn reduced the local heat dissipation. Consequently, thermal runaway in MOS transistors does not occur in contrast to BJTs.

With respect to high frequency operation, high switching speeds can be achieved in smart power ICs, reducing the size of the passive devices of the circuit. The switching frequency is limited by the parasitic capacitances. These properties come at a cost. Compared to bipolar devices, higher static power losses occur due to the limited unipolar current conduction and the lack of charge modulation in the ohmic drift region [19]. Therefore, improvement of LDMOS transistors in smart-power ICs is dominated by the reduction of on-state device resistance.

5.2.1 *Low Drift Region Resistance with "Reduced Surface Field"*

Early power semiconductor devices, mainly bipolar junction transistors and diodes, were fabricated using a lightly doped drift region constituting to a pin-type device. The requirement of using low doping concentrations (compared to typical drain doping concentrations) in order to achieve high blocking voltages brought the drawback of high on-state resistance in unipolar devices. Therefore, device technologists have been striving to increase the doping concentration of this drift region without degrading breakdown characteristics. It is for these considerations that the Reduced Surface Field effect was discovered.

5.2.1.1 Single Acting RESURF Topology

When the first integrated DMOS transistor for microwave applications was reported, its unexpectedly high breakdown voltage of 300 V could not be explained [12]. Another 7 years later, the principle of vertical charge compensation in lateral power devices which was named after reduced surface field (RESURF) was experimentally determined and described [13]. Considering the lateral pin-diode structure shown in Fig. 5.2 which consists of a deep n-well on a lightly p-doped substrate, the RESURF principle is illustrated.

When a positive potential to the n-contact is applied, a space-charge region extends from the pn-junction into the n-well. This resembles the blocking operation of the pin-diode. The electric field distributions at the surface of the device and at

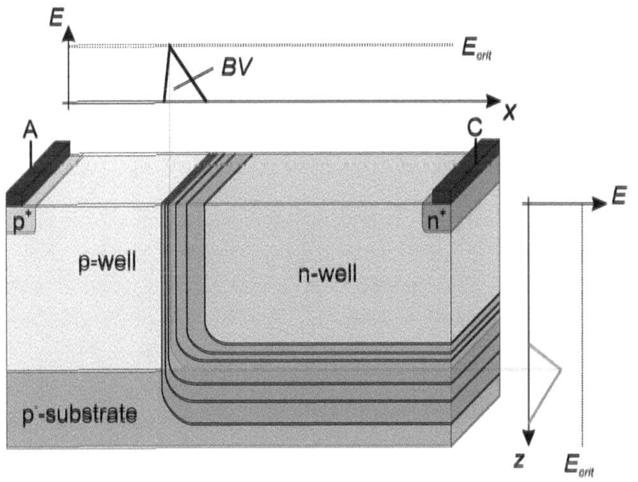

Fig. 5.2 Lateral pin-diode with *deep* n-well: structure and electric fields at avalanche breakdown—space charge region and electric field lines are simplified

the lateral n-well/p-substrate junction are also schematically depicted in Fig. 5.2. Due to the higher doping concentration of the n-well compared to the p-substrate, the space charge regions extends further in the p-substrate than in the n-well. The highest electric field develops at the vertical p^+/n-well junction. Only a low breakdown voltage—based on the doping concentration of the n-well can be applied to this pin-structure, and avalanche breakdown initiates near the surface of the device. This effect is particularly enhanced because surface states (interface states, fixed oxide charges) may locally increase the electric field at the device surface [20].

Note that the lateral n-well/p-substrate junction by itself can provide a higher breakdown voltage than the vertical p^+/n-well junction due to the lower doping concentration of the p-substrate. By reducing the thickness of the n-well sufficiently (see Fig. 5.3), an influence of this shallow lateral n-well/p-substrate junction on breakdown voltage becomes imminent.

Now, the n-well is fully depleted of free charge carriers by the lateral n-well/p-substrate junction before the critical electric field develops across the vertical p^+/n-well junction. In fact, a more or less constant lateral electric field develops at the device surface which is much lower than for the thick n-well (hence the name RESURF). Now, the breakdown voltage is entirely determined by the lateral junction within the semiconductor material away from its surface. Avalanche breakdown will be initiated at the cylindrical corner of this junction where electric field crowding occurs. The inevitable peaks in the lateral electric field depicted in Fig. 5.3 originate from the presence of the n^+- and p^+-regions that result in a localized lateral charge removal at the edges of the drift region [21].

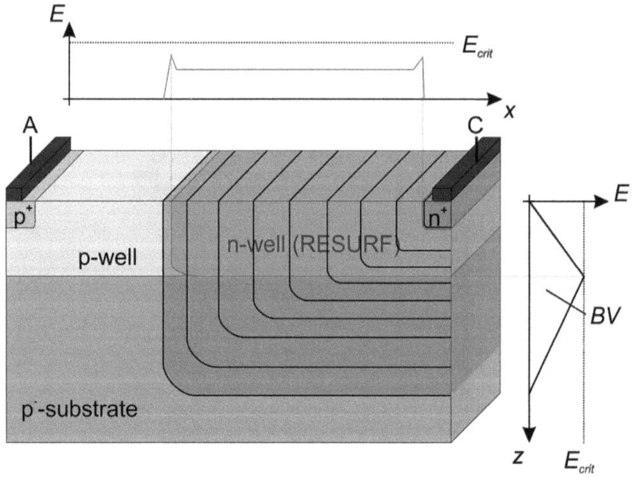

Fig. 5.3 Lateral pin-diode with *shallow* n-well and similar doping concentration as in Fig. 5.2: structure and electric fields at avalanche breakdown—space charge region and electric field lines are simplified

In summary, the RESURF principle relies on the removal of all free charge carriers by the extension of the lateral pn-junction before the vertical pn-junction goes into avalanche conditions. The doping concentration of the n-well can be increased independent of the breakdown voltage as long as all charges can be removed by the lateral junction. Since a high doping concentration results in a smaller space-charge region at a given voltage, a higher doping concentration necessitates the fabrication of a thinner layer. In order to investigate the impact of the RESURF design on on-state resistance, the dependence on the design parameters are reviewed next. For an n-doped RESURF region with homogeneous doping concentration $N_{n\text{-}well}$ with depth $d_{n\text{-}well}$, any space charge extending across a pn-junction gives rise to an electric field according to Poisson's equation:

$$Q = \varepsilon_0 \varepsilon_r E_{max} \cdot \sqrt{\frac{N_{p-sub}}{N_{p-sub} + N_{n-well}}} = q N_{n-well} d_{n-well}. \tag{5.1}$$

This equation yields a value for the product of doping concentration of the n-well and its depth (the *RESURF dose*) that must be present in the n-well region [21]:

$$N_{n-well} d_{n-well} = \frac{\varepsilon_0 \varepsilon_r E_{max}}{q} \cdot \sqrt{\frac{N_{p-sub}}{N_{p-sub} + N_{n-well}}}. \tag{5.2}$$

To prevent the electric field from reaching the critical value for avalanche breakdown at the pn-junction, this charge must be limited. Using silicon processing technology, i.e. ion implantation, the n-well region will be at least equal if not more highly doped than the p-substrate for reliable device fabrication. That dictates that the RESURF dose must be below

$$N_{n-well} d_{n-well} \leq \frac{\varepsilon_0 \varepsilon_r E_{max}}{q} \cdot \frac{1}{2} \sqrt{2} \tag{5.3}$$

in order to prevent premature device breakdown [22]. As a rule of thumb for silicon, the RESURF dose is typically in the range of $1 \times 10^{12}/\text{cm}^2$ [21]. Regarding the application of this dose, a Gaussian distribution of this charge within the vertical direction of the n-well as it is achieved by ion implantation and subsequent diffusion yields a good approximation of a uniformly doped region. Therefore, this value constitutes to an implantation dose typically used in silicon devices to fabricate RESURF regions and junction termination extensions. Depending on the interface state density and fixed oxide charges at the surface of the semiconductor, the value may differ between different fabrication technologies. Basically, a wide range of doping concentrations and n-well depths would be possible. Following the relation in (5.3), this is shown in Fig. 5.4 along with a technologically relevant process window.

Significant deviation from this dose will result in significant reduction of the breakdown voltage. This is true, in particular, if the RESURF dose is too high.

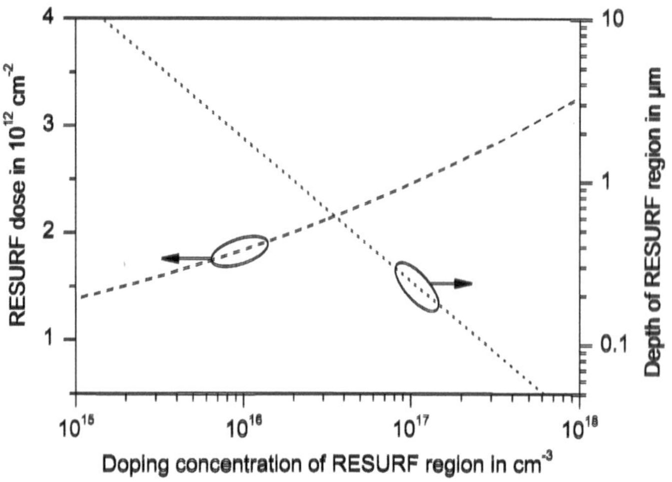

Fig. 5.4 Doping concentration and depth of the RESURF region satisfying the RESURF condition

Then, the n-well cannot be fully depleted vertically before the lateral electric field reaches the critical value for avalanche breakdown. Consequently, the RESURF region acts as a drain extension at the surface of the semiconductor with a small curvature at the channel side. For too low a dose, the vertical electric field becomes limited by the smaller amount of charges in the n-well. Moreover, a pronounced electric field peak arises at the n^{+}-contact region [13]. Finally, the p-substrate/n-well junction can be subject of a premature breakdown if the substrate doping concentration is too low. These effects also limit the breakdown voltage. When modelling the effect of the RESURF region, the gate field plate and the cylindrical curvature of the p-body have to be considered. O'Neil and Alonas [23] derived an analytical model to account for these effects in LDMOS transistors without RESURF design. Based on that work, an analytical approach to the optimization of RESURF LDMOS transistors has been presented by Parpia and Salama [24] in 1990. The impact of the RESURF dose on the breakdown voltage of the RESURF design can be estimated without 2D TCAD simulations as presented in Fig. 5.5.

From Fig. 5.6 it is apparent that this lateral pin-structure is the fundamental structure of planar LDMOS devices where the n-well of the underlying pin-diode acts as the drift region.

Therefore, RESURF designs can be directly applied to these devices. In turn, this allows for a significant reduction of the drift region resistance because a higher doping concentration for the drift region can be used. From the approximation of drift region resistance

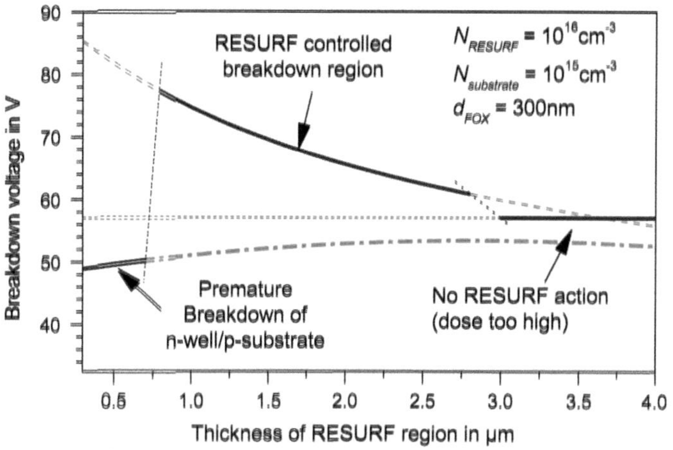

Fig. 5.5 Breakdown voltage depending on thickness of the RESURF region

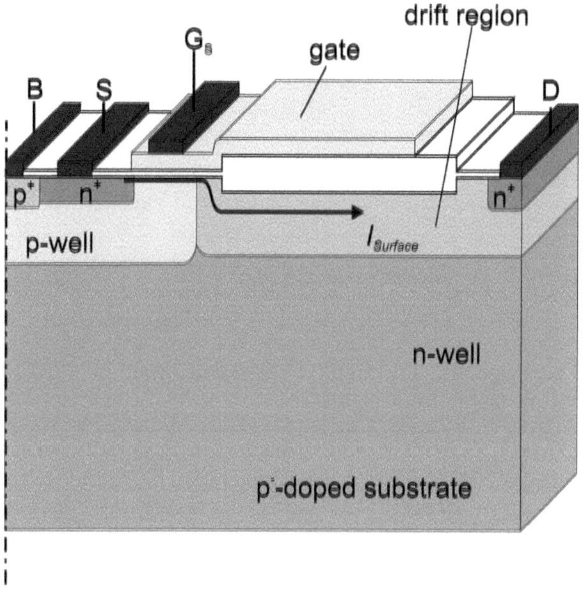

Fig. 5.6 Junction-isolated LDMOS device with RESURF design and gate field plate

$$R_{drift,RESURF} = \frac{1}{q\mu_{n,drift}N_{drift,RESURF}} \cdot \frac{L_{drift}}{d_{drift} \cdot W_{Drift}}, \tag{5.4}$$

and using (5.1) and (5.3) for a manufacturable RESURF design in modern CMOS technology results in an expression which only depends on design properties of the device and material constants of the semiconductor material.

$$R_{drift,RESURF} \geq \frac{1}{\mu_{n,drift}\varepsilon_0\varepsilon_r E_{crit}} \cdot \sqrt{2} \cdot \frac{L_{drift}}{W_{Drift}} \tag{5.5}$$

The depth of the RESURF-based drift region is determined by the doping concentration and the material dependent physical parameters. Calculating the area independent drift region resistance from (5.5) expanding by

$$A = L_{Drift} \cdot W_{Drift}, \tag{5.6}$$

and including the approximation of a constant lateral electric field in the drift region.

$$L_{drift} = \frac{BV}{E_{crit}}, \tag{5.7}$$

yields a purely material dependent relationship between the drift region resistance and the breakdown voltage:

$$R_{drift,RESURF} \cdot A \geq \frac{1}{\mu_{n,drift}\varepsilon_0\varepsilon_r E_{crit}^3} \cdot \sqrt{2} \cdot BV^2. \tag{5.8}$$

Using Fulop's approximation for avalanche breakdown in silicon [25], the relationship between breakdown voltage and critical electrical field is given by

$$\frac{E_{crit,Si}}{\mathrm{V\,cm^{-1}}} = 7.95 \cdot 10^5 \cdot \left(\frac{BV}{V}\right)^{-\frac{1}{6}}. \tag{5.9}$$

Hence, the area-specific drift region resistance limit of a RESURF-LDMOS transistor in CMOS technology yields

$$\frac{R_{drift,RESURF} \cdot A}{\Omega\,\mathrm{cm}^2} \geq 2.03 \cdot 10^{-9} \cdot \left(\frac{BV}{V}\right)^{2.5}. \tag{5.10}$$

The theoretical limit of drift region resistance versus breakdown voltage from (5.10), (4.16) and (4.25) are depicted in Fig. 5.7.

With an optimum choice of the doping concentrations and for breakdown voltages above 50 V, this constitutes to a significant improvement over the $R_{DS,on}$ versus BV Figure-of-Merit for the pin-type drift region as presented in Chap. 4. Conversely, a RESURF design for a targeted breakdown voltage below 30 V will result in a higher drift region resistance than for an extended-drain LDMOS.

Here, drift region resistance does not depend on the depth of the RESURF region which is inherently linked to the doping concentration by (5.1). This is also in

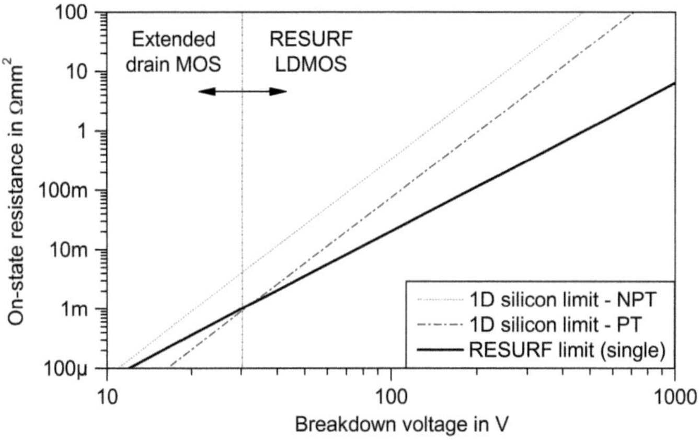

Fig. 5.7 Trade-off between drift region resistance and breakdown voltage in single-acting RESURF-based LDMOS transistor

contrast to the lateral 1D silicon limit as deduced in Chap. 4, endowing the Figure-of-Merit for RESURF-based LDMOS transistors with a universal validity across all single RESURF designs that only depends on the desired blocking voltage.

Regarding the considerations of static power losses discussed in Chap. 4 for LDMOS transistors utilizing pin-type drift region designs in switch-mode power supplies, a relationship between power losses and breakdown voltage can be derived similar to (4.31). For the RESURF-based LDMOS transistors, it holds that

$$P_{loss,on} = 2.03 \cdot 10^{-9} \cdot \left(\frac{BV}{V}\right)^{2.5} \cdot \frac{P_{trans}^2}{k_s^2 \cdot BV^2 \cdot A \cdot \delta^2} \sim BV^{\frac{1}{2}}. \qquad (5.11)$$

Therefore, it is desirable to increase the breakdown voltage to reduce static power losses for a given power density. The benefit of reducing the current density in the transistor is more beneficial than the increase in device resistance due to a high operating voltage.

5.2.1.2 Multi-acting RESURF Designs

To further improve the $R_{DS,on}$ versus BV Figure-of-Merit, multi-acting RESURF designs have been introduced. Such designs facilitate additional p-doped regions within the semiconductor to deplete the drift region from different areas. A decent summary of different RESURF topologies is given by Ludikhuize [21]. The topology of a double-acting RESURF design is presented in Fig. 5.8.

Here, the drift region is not only depleted from the substrate, but an additional p-region situated directly below the field oxide also depletes the drift region from above. Again, careful choice of the doping concentrations allows for an effective

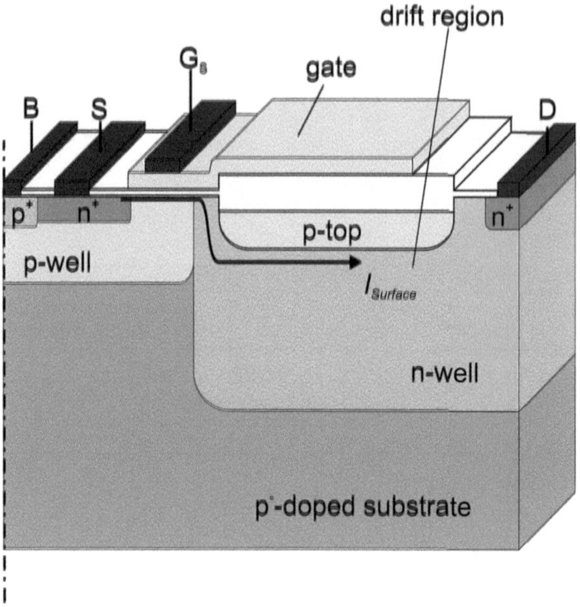

Fig. 5.8 LDMOS device with double-acting RESURF design, adopted from [26]

doubling of the RESURF dose in the drift region according to the relationship of the RESURF dose of the double-acting RESURF [22].

$$N_{drift}d_{drift} = \frac{\varepsilon_0\varepsilon_r E_{max}}{q}\cdot\left(\sqrt{\frac{N_{n-\text{well}}}{N_{p-\text{top}}+N_{n-\text{well}}}}+\sqrt{\frac{N_{p-\text{sub}}\cdot N_{n-\text{well}}}{N_{p-\text{top}}\cdot\left(N_{p-\text{sub}}+N_{n-\text{well}}\right)}}\right). \tag{5.12}$$

Again, setting an upper boundary for the RESURF dose from a process technology point of view dictates that

$$N_{p-\text{top}} \geq N_{n-\text{well}} \geq N_{p-\text{sub}} \tag{5.13}$$

Consequently, the relationship between drift region resistance and breakdown voltage that can be achieved using a double-acting RESURF becomes

$$\frac{R_{drift,D-RESURF}\cdot A}{\Omega\,\text{cm}^2} \geq \frac{1}{q\mu_{n,drift}\varepsilon_0\varepsilon_r E_{crit}^3}\cdot\frac{\sqrt{2}}{2}\cdot BV^2 = 1.02\cdot 10^{-9}\cdot\left(\frac{BV}{V}\right)^{2.5}. \tag{5.14}$$

Basically, this topology can be extended to an even larger degree of multi-level RESURF integration (e.g. [26]), but device processing becomes more complex and expensive. In the context of LDMOS devices in integrated circuits for operation voltages below 100 V, this is typically not feasible. The reason can be deduced from Fig. 5.9 which represents the trade-off between total device resistance and breakdown voltage for different CMOS-compatible LDMOS device designs, and considers standard non-isolated LDMOS transistors from the foundry processes listed in Table 5.1.

Again, it is evident from the strong deviation of the RESURF limit line that the drift region resistance does not significantly contribute to the total device resistance for devices with breakdown voltages well below 100 V. Here, the channel resistance becomes the dominating resistance distribution. The impact of channel

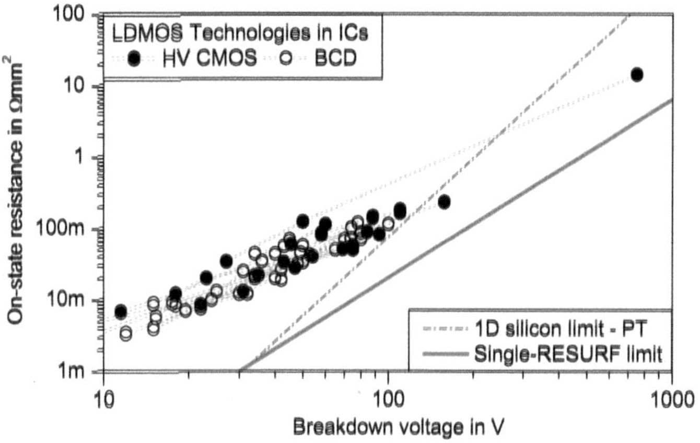

Fig. 5.9 Comparison between total device resistance and breakdown voltage for different LDMOS transistor technologies in commercially available smart-power ICs—theoretical 1D and RESURF limits for drift region resistance in silicon are given for comparison, data from sources given in Table 5.1

Table 5.1 Standard CMOS- and BiCMOS-based foundry processes with integrated LDMOS transistors reported in literature

Company	Type	Technology node in nm	Source
TSMC	BCD	250	[27]
Texas Instruments	BCD	250	[28]
ST Micro	BCD	180	[32]
Infineon	BCD	130	[34]
MagnaChip	BCD	180	[35]
Toshiba	BCD /HV-CMOS	130	[33]
DongBu	BCD /HV-CMOS	180	[36, 37, 38]
X-FAB	HV-CMOS	180/350	[29, 30]
IBM/AMS	HV-CMOS	180	[31]
ON Semi (AMI Semi)	HV-CMOS	350	[39]
Jazz Semi	HV-CMOS	180	[40]
Motorola	HV-CMOS	250	[41]

resistance will be reviewed in Chap. 7 in detail. Additionally, current crowding at the individual current paths through the drift region becomes more severe with each additional p-region situated in the drift region. The data shown in Fig. 5.9 will be used through the textbook to compare the research results on novel device concepts to the state-of-the-art RESURF LDMOS transistors.

5.2.2 Quasi-vertical Power Devices and Deep Trench Isolation in Integrated Circuits

A higher current density in integrated power transistors can also be achieved using quasi-vertical device structures, i.e. the application of a vertical drift region and collection of the drain current at the bottom of the device in a highly doped redistribution layer from where it can flow along a n^+-sinker back to the surface of the semiconductor. These sinkers are fabricated using deep trench technology [39]. This concept is also widely used for bipolar power transistors in bipolar and BiCMOS technologies. For quasi-vertical LDMOS devices trench pattern technologies are enabling new designs, and the combination of trench gates and quasi-vertical transistors will be discussed in Chap. 7. Deep trench technology is also combined with planar LDMOS topology as an isolation technology to increase capability under high side operation and to reduce leakage currents [42].

5.2.3 Field Plates

Another technique that is necessary for formidable device performance of LDMOS transistors (particularly at operating voltages beyond 30 V when the device is operated in the linear region) is the incorporation of field plates. Looking at a LDMOS design without field plate as shown in Fig. 5.10, a high electric field at the drain side of the channel is observed.

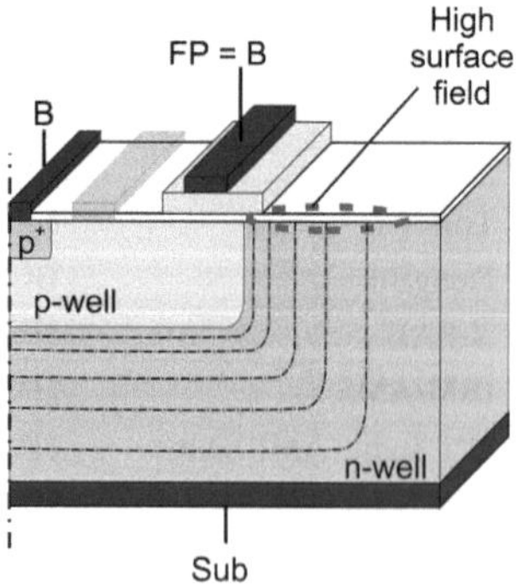

Fig. 5.10 Electric field distribution in a simplified LDMOS transistor cell without field plate

Operating this transistor can result in hot carrier injection (electrons) into the drain sided portion of the gate dielectric, causing threshold voltage shift and local degradation of the dielectric over time [43]. In the worst case, the transistor deviates from its initial parameters beyond the scope of the specification. In order to limit the electric field at the drain side of the channel region, a field plate can be introduced above the field oxide of the drift region adjacent to the channel region (see Fig. 5.11).

Depending on the bias of the field plate, the electric field at the drain-side of the channel region can be intensified or repelled. In order to protect the channel region from high electric fields, a bias negative compared to the bias at the drain side of the channel region must be imposed.

Moreover, the field plate is also capable of shielding the RESURF region from the overlying dielectric layers, passivation layers and mold compounds [43]. All these layers are potential sources of additional charges which could deteriorate the RESURF design [44, 45].

Conveniently, a gate voltage of 0 V is usually applied to LDMOS transistors in off-state. If switching for RF applications is not required, an established method of applying said bias to the field plate is to connect the field plate with the gate electrode to form a *gate field plate* as depicted in Fig. 5.12.

The gate field plate can be easily realized by extending the polysilicon of the gate electrode itself to the desired length of the field plate, which also does not require additional fabrication effort. The bias on the field plate is then linked to the semiconductor region below the field oxide and injection of channel electrons as hot carriers into the dielectric is minimized. However, the gate field plate leads to an increase in gate–drain capacitance due to the large gate–drain overlap across the field oxide:

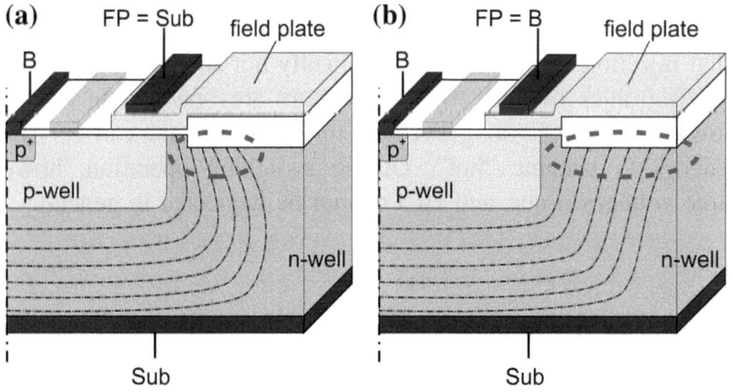

Fig. 5.11 Electrostatic potential distribution near channel region of a simplified LDMOS transistor cell with a field plate located on the field oxide next to the channel region with field plate potential tied to **a** substrate (i.e. drain potential of LDMOS) and **b** body region (i.e. source potential of LDMOS)

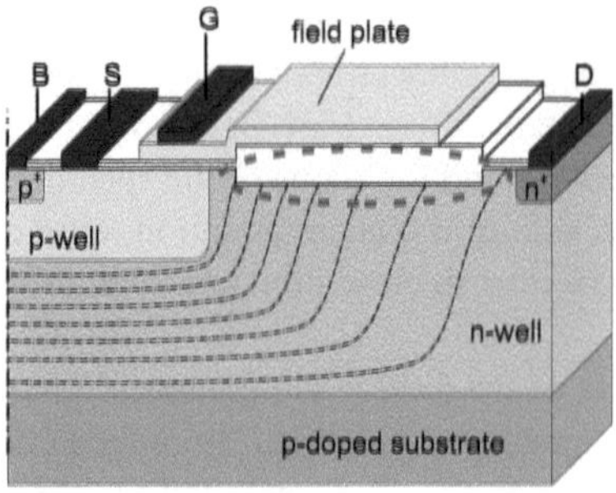

Fig. 5.12 Electric field distribution in a junction isolated LDMOS transistor with a field plate that is tied to the gate electrode under blocking condition

$$C_{gd,fp} = \frac{2L_{fp}}{L_{cell}} \left(\frac{C_{fox} C_{Sub,drift}}{C_{fox} + C_{Sub,drift}} \right) \tag{5.15}$$

with

$$C_{fox} = \frac{\varepsilon_0 \varepsilon_r}{d_{fox}} \tag{5.16}$$

With the definitions of LDMOS device capacitances introduced in Chap. 4, the gate–source, gate–drain and drain–source capacitances for an LDMOS transistor with a gate field plate are as follows:

$$C_{gs} = C_{gs,o} + C_{gb,o} \tag{5.17}$$

$$C_{gd} = C_{gd,o} + C_{gd,fp} \tag{5.18}$$

$$C_{ds} = C_{ds,scr} \tag{5.19}$$

Note that injection of hot carriers is typically not severe in LDMOS transistors for power electronics because these transistors are operated in linear operation mode at low drain–source voltages. Under this condition, the carriers cannot obtain enough energy to become "hot". During switching operation, however, high drain–source voltages occur, and HCI cannot be neglected in general.

5.2.4 High-Side Switching Capability

In order to fabricate LDMOS transistors in integrated circuits for switch-mode circuits using half- or full-bridge topologies, some of the transistors must be isolated from the substrate. The p-doped substrate usually comprises the most negative operating potential of the circuit. The design shown in Fig. 5.12 includes a junction

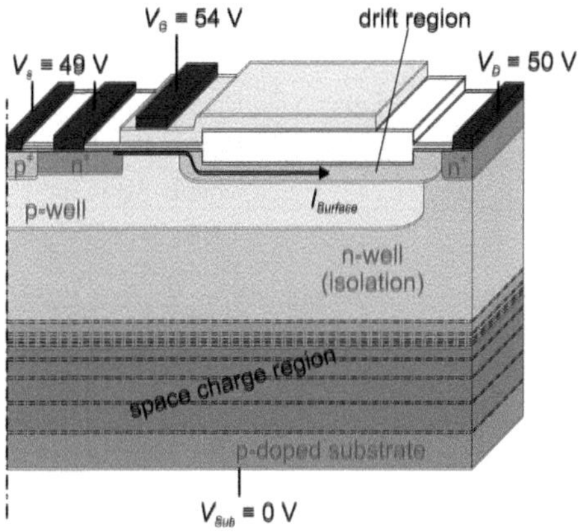

Fig. 5.13 Electric field distribution and space charge region in LDMOS transistor with junction isolation to substrate at on-state when operated as a high-side switch; simplified drawing where p-substrate/n-well junction is highlighted

isolation of the substrate by implementing a pn-junction between the drain and the substrate, i.e. isolating the n-drift region. When a high drain voltage is applied, a space-charge region extends from substrate to the n-well as depicted in Fig. 5.13.

Here, the drain electrode can be safely biased to the supply voltage of the circuit while the source potential can float anywhere between the supply voltage and the voltage of the substrate.

5.2.5 Implementation of RESURF-Based Technologies in Smart Power IC Processing Technologies

Several device concepts based on multiple RESURF and optimization of the gate field plate design has been reported over the last decades. The electrical properties from these works relevant for power electronic applications are given in Table 5.2 for fabricated devices and in Table 5.3 for simulation studies.

This data represents some of the major work that has been carried out on the development of RESURF-based LDMOS transistors over the last 2 decades. A direct comparison between different designs is difficult because there are no standardized operating voltages across different foundries. Each company provides an individual maximum voltage rating for its LDMOS transistors. In an effort to compare different transistor designs, the breakdown voltage has to be considered. However, it should be noted that the breakdown voltages of different LDMOS transistors rated for the same maximum operating voltage may be different due to restrictions arising from other reliability limiting effects, e.g. hot carrier injection or Kirk effects.

These RESURF-based technologies have been widely augmented in commercial smart power IC technologies as is conceivable from Fig. 5.14.

Table 5.2 Breakdown voltage, on-state resistance and gate charge measured from fabricated LDMOS transistors using RESURF design reported in literature

BV (V)	$R_{DS,on}$ (Ω mm^2)	Q_{gd} (nC/mm^2)	Device type	Substrate	Source
740	20	–	Double-RESURF	Silicon	[43]
750	19	–	Double-RESURF	Silicon	[46]
800	16	–	Triple-RESURF	Silicon	[47]
400	4	–	Triple-RESURF	Silicon	[47]
190	0.5	–	Double-RESURF	SIO	[48]
35	0.27	–	Single-RESURF	Silicon	[49]
700	20	–	Single-RESURF	Silicon	[49]
1100	48	–	Single-RESURF	Silicon	[49]
28	0.0177	–	Single RESURF	Silicon	[50]
28	0.0187	–	EDMOS	Silicon	[50]
35.2	0.0429	40	EDMOS	Silicon	[51]
47	0.028	–	FRESURF	Silicon	[52]
62	0.038	–	FRESURF	Silicon	[52]
93	0.082	–	FRESURF	Silicon	[52]
33	0.025	–	EDMOS	Silicon	[53]
38	0.0225	–	EDMOS	Silicon	[40]
25	0.025	–	ISO EDMOS	Silicon	[40]
22	0.0396	0.308	RESURF + Sinker	Silicon	[54]
36	0.077	0.174	RESURF + Sinker	Silicon	[54]
100	0.18	–	RESURF	Silicon	[55]
55	0.05	–	RESURF	Silicon	[56]
97	0.2	–	RESURF	Silicon	[57]
28	0.024	–	RESURF	Silicon	[58]

Table 5.3 Breakdown voltage, on-state resistance and gate charge from simulated LDMOS transistors implementing RESURF topology reported in literature

BV (V)	$R_{DS,on}$ (Ω mm^2)	Q_{gd} (nC/mm^2)	Device type	Substrate	Source
800	28	–	Double-RESURF	Silicon	[59]
83	0.21	–	Single-RESURF	SOI	[60]
91	0.145	0.882	Double-RESURF	Silicon	[61]
153	0.45	1.51	RESURF	SOI	[62]
48.1	0.08	0.377	Multi-Gate	SOI	[63]
29.6	0.13	1.006	Single-Gate	SOI	[63]

5.3 LDMOS Transistors for Radio-Frequency Applications

For RF amplifiers, high operating frequencies, gain and linearity are desired. The RF LDMOS transistors discussed in this chapter excel over silicon bipolar transistors in monolithic microwave integrated circuits (MMICs) for RF applications.

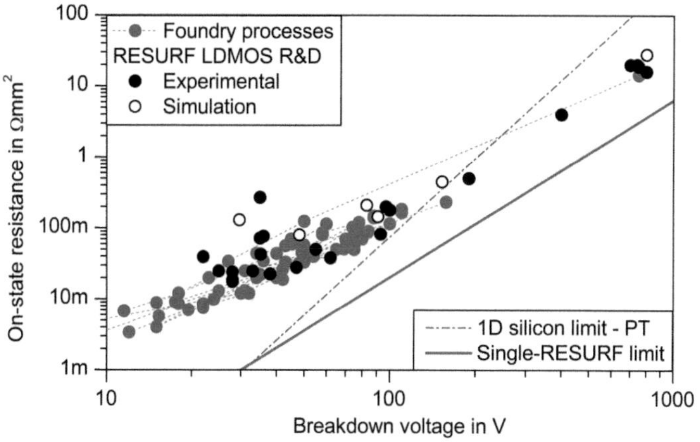

Fig. 5.14 Comparison of LDMOS developments based on multiple RESURF, drain-extension and gate field plate optimizations with state-of-the-art device properties, data from Tables 5.2 and 5.3

Switching frequency in LDMOS transistors is limited by device capacitances, and the key effort in improving RF performance is to minimize these capacitances. Additionally, a reduction of device resistance increases the efficiency of these devices. Beside the reduction in feature size following More-Moore integration, additional concepts have been implemented to increase device performance.

5.3.1 Source Sinker

In order to minimize the source impedance for high frequency operation—see (4.54)—a low ohmic source sinker has been introduced for stand-alone RF LDMOS transistors and those integrated into MMICs. The cross-section of an RF LDMOS transistor with a p^+-sinker is illustrated in Fig. 5.15.

Hereby, the source at the surface of the planar LDMOS device can be connected to the backside of the semiconductor. This offers the advantage of mounting the device backside to the circuit board eliminating the source impedance typically encountered when bond wiring is used to assemble lateral devices. Fabrication of the p^+-sinker is routinely performed by ion implantation and a long diffusion process. Due to the high thermal budget, this process step is usually performed before any other doping processes [17].

5.3.2 Ground Shield

Performance of LDMOS transistors for radio-frequency operation can be enhanced by the minimization of the gate–drain capacitance as demonstrated in Chap. 4. The

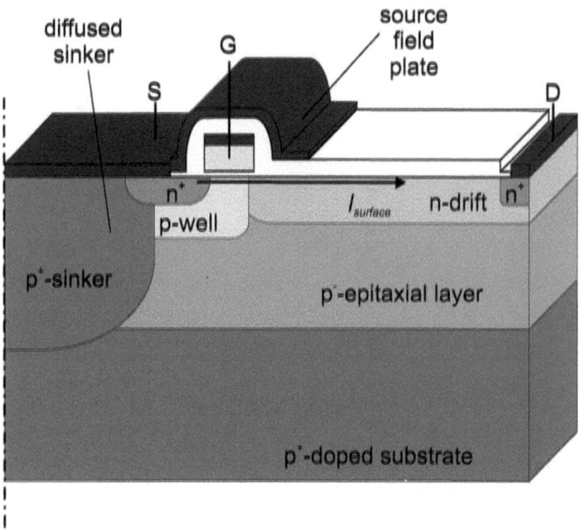

Fig. 5.15 RF LDMOS transistor with source field plate and p$^+$-sinker

impact of capacitances in (4.52) dictates that the transition frequency (and the gain-bandwidth product of the amplifier circuit) can be increased, when part of the gate–drain capacitance is "converted" to a gate–source capacitance. This can be achieved by amending the field plate technique from Fig. 5.12. Instead of tying the field plate to gate potential, where it contributes to the gate–drain capacitance, the field plate is tied to source potential as illustrated in Fig. 5.16 forming a *source field plate*, *faraday shield* or *ground shield*.

Under this stipulation, the capacitance between the field plate and the drift region is charged by the drain and source electrodes and the gate is shielded (hence the name ground shield). Also, considering the gate–field plate capacitance C_{gfp} and the drain–field plate capacitance C_{dfp}, the device capacitances can be estimated using the considerations from Chap. 4.

$$C_{gs} = C_{gs0} + C_{gb0} + C_{gfp}. \tag{5.20}$$

$$C_{gd} = C_{gd0} + \frac{C_{gfp} \cdot C_{dfp}}{C_{gfp} + C_{dfp}}. \tag{5.21}$$

$$C_{ds} = C_{ds0} + C_{dfp}. \tag{5.22}$$

While the gate–source and drain–source capacitances increase due to the presence of the source field plate, the gate–drain capacitance can be minimized. Here, the largest portion of the drain voltage is coupled to the gate across the series combination of C_{gfp} and C_{dfp}, i.e. across the source field plate.

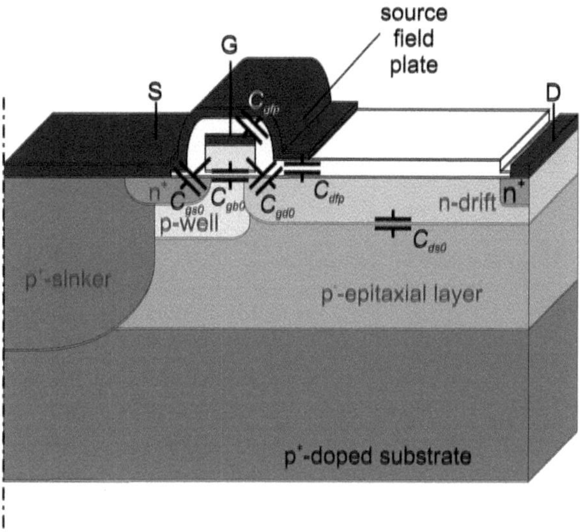

Fig. 5.16 Parasitic device capacitances in an LDMOS transistor with a field plate that is tied to source potential

Given that the field plate remains negatively biased compared to the drain voltage, the channel region remains shielded from the high electric fields in the drift region. Therefore, injection of hot carriers into the gate dielectric can also be minimized. This is of particular interest for RF LDMOS transistors where operating in saturation (at high drain–source voltages) is common-place, and a significant reduction of gate–drain capacitance by tying the field plate to source potential was demonstrated [64]. In contrast to the polysilicon field plate, fabrication of the ground shield requires additional processing effort in the interconnection layers.

5.3.3 *Design Considerations for RF Transistors in Integrated Circuits*

With continuous reduction of feature size in highly integrated BiCMOS and CMOS technologies, the device capacitances have become very small. This has led to integration of RF LDMOS transistors into application specific ICs. With the gate–drain-capacitance being the most limiting parasitic, the concept of the ground field plate has been incorporated into these designs. The source sinker, however, is not requires as the source of the RF transistors is directly connected to the adjacent circuitry on-chip, and inductance for RF applications can be neglected. The performance of different RF LDMOS transistors integrated into CMOS and BiCMOS processes are summarized in Table 5.4 and maximum oscillation frequency for different breakdown voltages are displayed in Fig. 5.17.

Table 5.4 Breakdown voltage, cut-off and maximum oscillation frequencies for RF transistors using extended drain (EDMOS) or RESURF (LDMOS) topologies reported in literature

BV (V)	f_T (GHz)	f_{max} (GHz)	Device type	Substrate	Source
18	–	30	RF EDMOS	Silicon	[40]
25	–	18	RF EDMOS	Silicon	[40]
65	–	1.2	RF LDMOS	Silicon	[65]
27	21	55	RF EDMOS	Silicon	[66]
14.8	15.2	25.1	RF EDMOS	SOI	[67]
20	25	55	RF EDMOS	Silicon	[68]
24	24	53	RF EDMOS	Silicon	[69]
16.5	22	29	RF EDMOS	Silicon	[70]
16.5	25	40	RF EDMOS	Silicon	[71]
13.5	21.2	39.3	RF EDMOS	Silicon	[72]
15.5	31	47	RF LDMOS	Silicon	[73]
15	30	52	RF EDMOS	Silicon	[74]
26.5	17	43	RF EDMOS	Silicon	[74]
60	15	38	RF LDMOS	Silicon	[75]

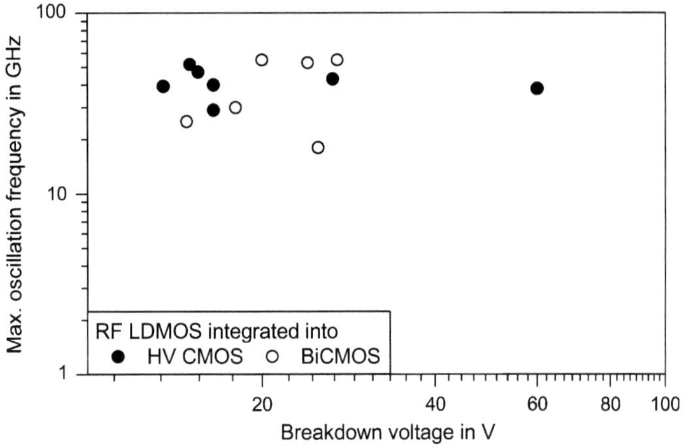

Fig. 5.17 Breakdown voltage versus maximum oscillation frequency for RF LDMOS transistors integrated into (Bi)CMOS technology, data from Table 5.4

These results will be used as a benchmark for RF LDMOS transistors implemented using the novel device concepts described throughout the rest of this textbook.

5.3.4 LDMOS Transistors in Silicon-on-Insulator Technology

In order to improve device performance of CMOS circuits for low voltage, low power and high frequency applications, silicon-on-insulator (SOI) technology has been introduced [76]. An extensive summary dedicated to fabrication technologies for SOI wafers is given by Celler [77]. Dielectric isolation for integrated circuits is employed instead of junction isolation. Thereby, the capacitances between the individual devices and the substrate and leakage currents can be reduced. Moreover, parasitic bipolar transistors are avoided and latch-up effects which can limit switching operating are prevented. The dielectric isolation also incurs less leakage current than junction isolation.

In dielectric isolated SOI devices, the bottom oxide acts as a lightly doped p-substrate for the n-doped drift region. It has been described by Merchant et al. [78] that very thin silicon layers on insulating substrates can be used to achieve very high breakdown voltages similar to the RESURF concept in junction-isolated LDMOS transistors. LDMOS transistors with breakdown voltage of 860 V were readily demonstrated [79]. These investigations were augmented by the development of analytical models to predict breakdown voltage of SOI-based power devices towards thinner silicon layers [80].

Subsequently, the field-plate and the RESURF concepts developed on junction-isolated LDMOS transistors were also carried over to SOI LDMOS transistors. In Fig. 5.18 the cross-section of a state-of-the-art LDMOS transistor is illustrated.

RF transistors fabricated on SOI using a 0.5 μm CMOS process were reported achieving maximum oscillation frequency of 25.1 GHz at a breakdown voltage of 14.8 V [67].

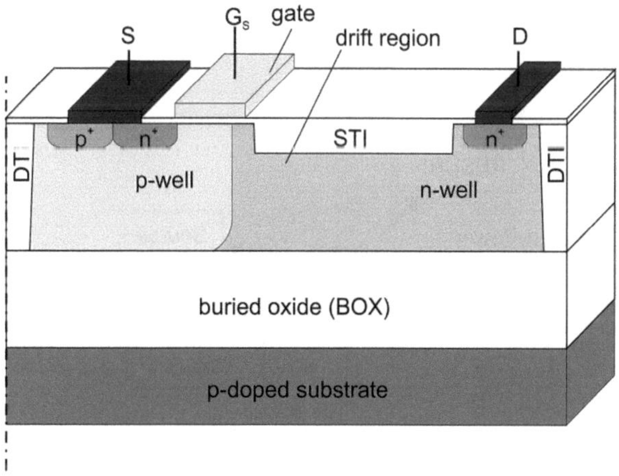

Fig. 5.18 Implementation of dielectrically isolated RF LDMOS with deep trench isolation (DTI) and shallow trench isolation (STI)

5.4 State-of-the-Art in Smart-Power IC and RF Amplifier Technologies

The invention of the DMOS transistor has led to a deployment of CMOS-based technology towards devices for high power applications. The introduction of the RESURF design constituted to a significant improvement in device performance over transistor designs with homogeneously doped drift regions. From there, silicon power devices were developed along two separate roads. High voltage devices are particularly desirable for energy conversion circuits, and the incorporation of gate field plates and the RESURF design were cornerstones for their success. High linearity and low device capacitances are imperative requirements for RF amplifiers in wireless broadcast applications with high data rates, and the implementation of source field plates enabled this. The RF LDMOS paved the way to high power RF solutions on silicon. The evolution of these two distinct device technologies (HV LDMOS and RF LDMOS paved the way to high power RF solutions on silicon. The evolution of these two distinct device technologies (HV LDMOS and RF LDMOS) is depicted in Fig. 5.19.

Smart-power ICs have evolved for high power densities at lower switching frequencies, whereas integrated RF amplifiers are characterized by high switching frequencies and lower power densities. Figure 5.19 will be used as a reference in Chap. 10 to depict the potential development of smart-power ICs and RF amplifiers based on the technologies and topologies reviewed in Chaps. 6–9.

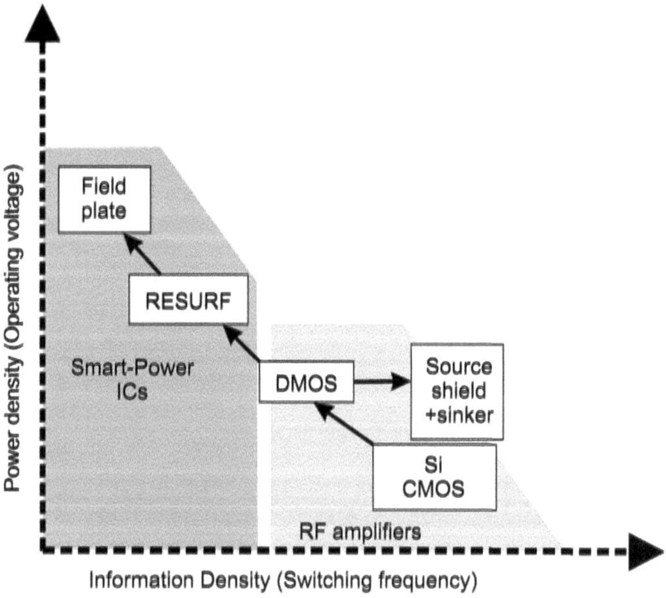

Fig. 5.19 Development of CMOS-based devices for smart-power and integrated RF amplifiers representing the state-of-the-art in silicon

References

1. J. Bardeen, W.H. Brattain, The transistor, a semi-conductor triode. Phys. Rev. **74**(2), 230–231 (1947)
2. W. Shockley, The theory of p–n junctions in semiconductors and p–n junction transistors. Bell Syst. Tech. J. **28**(4), 435–489 (1949)
3. W.J. Pietenpol, R.S. Ohl, Characteristics of Silicon Transistors, in *Conference on Electron Devices*, Ann Arbor, USA (1950)
4. G.K. Teal, E. Buehler, Growth of silicon single crystals and of single crystal p–n junctions. Phys. Rev. **87**, 190 (1952)
5. J.A. Hoerni, Planar silicon transistors and diodes. IRE Trans. Electron Devices **8**, 178 (1961)
6. R.N. Noyce, Semiconductor Device-and-Lead Structure. USA Patent 2,981,877, 25 April 1959
7. J.E. Lilienfeld, Method and Apparatus for Controlling Electric Circuits. USA Patent 1,745,175, 28 Jan 1930
8. O. Heil, Improvements in or Relating Amplifiers or Other Control Arrangements. UK Patent 439,457, 2 March 1934
9. M.M. Atalla, E. Tannenbaum, E.J. Scheibner, Stabilization of silicon surfaces by thermally grown oxides. Bell Labs Tech. J. **38**(3), 749–783 (1959)
10. D. Kahng, M.M. Atalla, Silicon–silicon dioxide field induced surface device, in *Solid State Device Research Conference*, Pittsburgh, USA 1960
11. W.F. Brinkman, D.E. Haggan, W.W. Troutman, A history of the invention of the transistor and where it will lead us. IEEE J. Solid-State Circuits **32**(12), 1858–1865 (1997)
12. H.J. Sigg, G.D. Vendelin, T.P. Cauge, J. Kocsis, D-MOS transistor for microwave applications. IEEE Trans. Electron Devices **9**(1), 45–53 (1972)
13. J.A. Appels, H.M.S. Vaes, High voltage thin layer devices, in *International Electron Devices Meeting*, Washington, USA, 1979
14. S. Colak, Effects of drift region parameters on the static properties of power LDMOST. IEEE Trans. Electron Devices **ED-28**(12), 1455–1466 (1981)
15. V.A.K. Temple, Junction termination extension (JTE), A new technique for increasing avalanche breakdown voltage and controlling surface electric fields in P-N junctions, in *International Electron Devices Meeting*, Washington, USA, 1977
16. O. Ishakawa, E. Hideya, A high-power high-gain VD-MOSFET operating at 900 MHz. IEEE Trans. Electron Devices **ED-34**(5), 1157–1162 (1987)
17. A. Wood, C. Dragon, W. Burger, High performance silicon LDMOS technology for 2 GHz RF power amplifier applications, in *International Electron Device Meeting*, San Francisco, USA, 1996
18. G. Deboy, M. März, J.-P. Stengl, H. Strack, J. Tihanyi, H. Weber, New generation of high voltage MOSFETs breaks the limit line of silicon, in *International Electron Device Meeting*, San Francisco, USA, 1998
19. S. Linder, *Power Semiconductors, Lausanne* (EFPL Press, Switzerland, 2006)
20. P. Moens, G. Van den bosch, D. Wojciechowski, F. Bauwens, H. De Vleeschouwer, F. De Pestel, Charge Trapping Effects and Interface State Generation in a 40 V Lateral Resurf pDMOS Transistor, in *European Solid State Device Research Conference*, Grenoble, France, 2005
21. A.W. Ludikhuize, A Review of RESURF Technology, *International Symposium on Power Semiconductor Devices & ICs*, pp. 11–18, 22–25 May 2000
22. M. Imam, Z. Hossain, M. Quddus, J. Adams, C. Hoggatt, T. Ishiguro, R. Nair, Design and optimization of double-RESURF high-voltage lateral devices for a manufacturable process. IEEE Trans. Electron Devices **50**(7), 1697–1701 (2003)
23. V.P. O'Neil, P.G. Alonas, Relation between oxide thickness and the breakdown voltage of a planar junction with field relief electrode. IEEE Trans. Electron Devices **ED-26**(7), 1098–1100 (1979)

24. Z. Parpia, C.A.T. Salama, Optimization of RESURF LDMOS transistors: an analytical approach. IEEE Trans. Electron Devices **37**(3), 789–796 (1990)
25. W. Fulop, Calculation of avalanche breakdwon of silicon p–n junctions. Solid-State Electron. **10**, 39–43 (1967)
26. M.M.-H. Iqbal, F. Udrea, E. Napoli, On the static performance of the RESURF LDMOSFETs for power ICs, in *International Symposium of Power Semiconductor Devices & ICs*, Barcelona, Spain, 2009
27. K. Chen, C. Chen, The foundry perspective on integrated power technologies, in *International Symposium on Semiconductor Devices & ICs*, Barcelona, Spain, 2009
28. S. Pendharkar, R. Pan, T. Tamura, B. Todd, T. Efland, 7 to 30 V state-of-the-art power device implementation in 0.25 μm LBC7 BiCMOS-DMOS process technology, in *International Symposium on Power Semiconductor Devices and ICs*, Kitayushu, Japan, 2004
29. D. Riccardi, A. Causio, I. Filippi, A. Paleari, L.V.A. Pregnolato, P. Galbiati, C. Contiero, BCD8 from 7 V to 70 V: a new 0.18 μm technology platform to address the evolution of applications towards smart power ICs with high logic contents, in *International Symposium on Power Semiconductor Devices and ICs*, Jeju, Korea, 2007
30. R. Rudolf, C. Wagner, L. O'Riain, K.-H. Gebhardt, B. Kuhn-Heinricht, B. von Ehrenwall, A. von Ehrenwall, M. Strasser, M. Stecher, U. Glaser, S. Aresu, P. Kuepper, A. Mayerhofer, Automotive 130 nm smart-power-technology including embedded flash functionality, in *International Symposium on Power Semiconductor Devices and ICs*, San Diego, CA, USA, 2011
31. Y.-S. Pang, Dependence of electrical characteristics of improved BCD nLDMOS on process and structural parameters. J. Korean Phys. Soc. **59**(2), 416–420 (2011)
32. K. Shirai, K. Yonemura, K. Watanabe, K. Kimura, Ultra-low on-resistance LDMOS implementation in 0.13 μm CD and BiCD process technologies for analog power IC's, in *International Symposium on Power Semiconductor Devices and ICs*, Barcelona, Spain, 2009
33. I.-Y. Park, Y.-K. Choi, K.-Y. Ko, C.-J. Yoon, B.-K. Jun, M.-Y. Kim, H.-C. Lim, N.-J. Kim und K.-D. Yoo, BD180—a new 0.18 μm BCD (Bipolar-CMOS-DMOS) technology from 7 V to 60 V, in *International Symposium on Power Semiconductor Devices and ICs*, Orlando, FL, USA, 2008
34. K.-Y. Ko, I.-L. Park, Choi Y-K, C.-J. Yoon, J.-H. Moon, K.-M. Park, H.-C. Lim, S.-Y. Park, N.-J. Kim, K.-D. Yoo, L. N. Hutter, BD180LV—0.18 μm BCD technology with best-in-class LDMOS from 7 V to 30 V, in *International Symposium on Power Semiconductor Devices and ICs*, Hiroshima, Japan, 2010
35. Y.-K. Choi, I.-Y. Park, H.-C. Lim, M.-Y. Kim, C.-J. Yoon, N.-J. Kim, K.-D. Yoo, L. N. Hutter, A versatile 30 V analog CMOS process in a 0.18 μm technology for power management applications, in *International Symposium on Power Semiconductor Devices and ICs*, San Diego, CA, USA, 2011
36. X-FAB, 0.18 micron modular analog mixed hv technology, http://www.xfab.com/de/technologien/cmos/. [Zugriff am 31st May 2014]
37. X-FAB, 0.35 Micron modular analog mixed signal technology with RF capability and HV extensions, http://www.xfab.com/de/technologien/cmos/. [Zugriff am 31st May 2014]
38. R. Minixhofer, N. Feilchenfeld, M. Knaipp, G. Röhrer, J. M. Park, M. Zierak, H. Enichlmair, M. Levy, B. Loeffler, D. Hershberger, F. Unterleitner, M. Gautsch, K. Chatty, Y. Shi, W. Posch, E. Seebacher, M. Schrems, J. Dunn, D. Harame, A 120 V 180 nm high voltage CMOS smart power technology for system-on-chip integration, in *International Symposium on Power Semiconductor Devices and ICs*, Hiroshima, Japan, 2010
39. F. De Pestel, P. Moens, H. Hakim, K. De Vleeschouwer, K. Reynders, T. Colpaert, P. Colson, P. Coppens, S. Boonen, D. Dolognesi, M. Tack, Development of a robust 50 V 0.35 μm based smart power technology using trench isolation, in *International Symposium on Power Semiconductor Devices and ICs*, Cambridge, UK, 2003
40. Z. Lee, R. Zwingman, J. Zheng, W. Cai, P. Hurwitz, M. Racanelli, A Modular 0.18 um analog/RFCMOS technology comprising 32 GHz FT RF-LDMOS and 40 V complementary

MOSFET devices, in *Bipolar/BiCMOS Circuits and Technology Meeting*, Maastricht, Netherlands, 2006
41. V. Parthasarathy, R. Zhu, V. Khemka, T. Roggenbauer, A. Bose, P. Hui, P. Rodriquez, J. Nivison, D. Collins, Z. Wu, I. Puchades, M. Butner, A 0.25 μm CMOS based 70 V smart power technology with deep trench for high-voltage isolation, in *International Electron Devices Meeting*, San Francisco, CA, USA, 2002
42. V. Parthasarathy, R. Zhu, V. Khemka, T. Roggenbauer, A. Bose, P. Hui, P. Rodriquez, J. Nivison, D. Collins, Z. Wu, I. Puchades, M. Butner, A 0.25 μm CMOS based 70 V smart power technology with deep trench for high-voltage isolation, in *International Electron Devices Meeting*, San Francisco, USA, 2002
43. Z. Hossain, T. Ishiguro, L. Tu, H. Corleto, F. Kuramae, R. Nair, Field-plate effects on the breakdown voltage of an integrated high-voltage LDMOS transistor, in *International Symposium on Power Semiconductor Devices and ICs*, Kitakyushu, Japan, 2004
44. A.S. Chen, A. Shafi, R.W. Busse, R. Orr, R.H. Y. Lo, Molding compounds for high breakdown voltage applications on power ic semiconductors, in *Proceedings of Electronic Components and Technology Conference*, San Jose, California, 1997
45. J.W. Osenbach, W.R. Knolle, Semi-insulating silicon nitride (SinSiN) as a resistive field shield. IEEE Trans. Electron Devices **37**(6), 1522–1528 (1990)
46. Z. Hossain, I. Mohamed, J. Fultonm, M. Tanaka, Double-resurf 700 V N-channel LDMOS with best-in-class on-resistance, in *International Symposium on Semiconductor Devices and ICs*, Santa Fe, NM, USA, 2002
47. D.R. Disney, A.K. Paul, M. Darwish, R. Basecki, V. Rumennik, A new 800 V lateral MOSFET with dual conduction paths, in *International Symposium on Power Semiconductor Devices and ICs*, Osaka, Japan, 2001
48. A.W. Ludikhuize, J.A. van der Pol, A. Heringa, A. Padiy, E.R. Ooms, P. van Kessel, G.J.J. Hessels, B. van Velzen, H. van der Vlist, J.H.H.A. Egbers, M. Stoutjesdijk, Extended (180 V) voltage in 0.6 um thin-layer silicon-on insulator BCD-technology on 1 um BOX for display, automotive and consumer applications, in *International Symposium on Power Semiconductors and ICs*, Santa Fe, USA, 2002
49. A.W. Ludikhuize, A versatile 700–1200 V IC process for analog and switching applications. IEEE Trans. Electron Devices **18**(7), 1582–1589 (1991)
50. K. Kinoshita, Y. Kawaguchi, T. Sano, A. Nakagawa, 20 V LDMOS optimized for high drain current condition—which is better, n-epi or p-epi?, in *International Symposium on Power Semiconductor Devices and ICs*, Toronto, Canada, 1999
51. H. Wang, H.P.E. Xu, W.T. Ng, K. Fukumoto, K. Abe, A. Ishikawa, Y. Furukawa, H. Imai, T. Naito, N. Sato, K. Sakai, S. Tamura, K. Takasuka, A novel orthogonal gate EDMOS transistor with improved dv/dt capability and figure of merit (FOM). IEEE Electron Device Lett. **29**(12), 1386–1388 (2008)
52. V. Khemka, V. Parthasarathy, R. Zhu, A. Bose, T. Roggenbauer, Floating RESURF (FRESURF) LDMOSFET devices with breakthrough BVdss-Rdson, in *International Symposium on Power Semiconductor Devices and ICs*, Kitakyushu, Japan, 2004
53. V. Parthasarathy, R. Zhu, W. Peterson, M. Zunino, R. Baird, A 33 V, 0.25 mOhm-cm^2 n-channel LDMOS in a 0.65 μm smart power technology for 20–30 V applications, in *International Symposium on Power Semiconductor Devices and ICs*, Kyoto, Japan, 1998
54. K. Sakamoto, M. Shiraishi, T. Iwasaki, Low on-resistance and low feedback charge, lateral power MOSFETs with multi-drain regions for high-efficient DC/DC converters, in *International Symposium on Power Semiconductor Devices and ICs*, Santa Fe, USA, 2002
55. S.T. Kong, P. Stribley, C. Lee, M. Ong, Integration of 100 V LDMOS devices in 0.35 μm CMOS technology, in *International Symposium on Power Semiconductor Devices and ICs*, San Diego, USA, 2011
56. N. Klein, S. Levin, G. Fleishon, S. Levy, A. Eyal und S. Shapira, Device Design Tradeoffs for 55 V LDMOS Driver Embedded in 0.18 Micron Platform, in *IEEE Convention of Electrical and Electronics Engineers in Israel*, Eilat, Israel, 2008

57. T. Efland, S. Malhi, W. Bailey, O.K. Kwon, W.T. Ng, M. Torreno, S. Keller, An optimized RESURF LDMOS power device module compatible with advanced logic processes, in *International Electron Devices Meeting*, San Francisco, USA, 1992
58. A.W. Ludikhuize, Self-aligned and shielded-RESURF LDMOS for dense 20 V power IC's, in *International Symposium on Power Semiconductor Devices and ICs*, Toronto, Canada, 1999
59. W. Chen, B. Zhang, Z. Li, A novel double RESURF LDMOS and a versatile JFET device used as internal power supply and current detector for SPIC. Microelectron. J. **37**(7), 574–578 (2006)
60. I. Cortes, P. Fernandez-Martinez, D. Flores, S. Hidalgo, J. Rebollo, Analysis of low-voltage super-junction LDMOS structures on thin-SOI substrates. Semiconduct. Sci. Technol. **23**, 015009 (2008)
61. T. Erlbacher, A.J. Bauer, L. Frey, Significant on-resistance reduction of LDMOS devices by intermitted trench gates integration. IEEE Trans. Electron Devices **59**, 3470–3476 (2012)
62. M.A. Amberetu, C.A.T. Salama, 150-V class superjunction power LDMOS transistor switch on SOI, in *International Symposium on Power Semiconductors and ICs*, Santa Fe, USA, 2002
63. M.J. Kumar, R. Sithanandam, Extended-p+ stepped gate LDMOS for improved performance. IEEE Trans. Electron Device **57**(7), 1719–1724 (2010)
64. I. Cortés, J. Roig, D. Flores, J. Urresti, S. Hidalgo, J. Rebollo, A numerical study of field plate configurations in RF SOI LDMOS transistors. Solid-State Electron. **50**, 155–163 (2006)
65. P.H. Wilson, A novel trench gate LDMOS for RF applications, in *International Crimean Conference on Microwave and Telecommunication Technology*, Sevastopol, Ukraine, 2003
66. N.R. Mohapatra, H. Ruecker, K.E. Ehwald, R. Sorge, R. Barth, P. Schley, D. Schmidt, H.E. Wulf, A complementary RF-LDMOS architecture compatible with 0.13 μm CMOS technology, in *International Symposium on Power Semiconductor Devices and ICs*, Naples, Italy, 2006
67. S. Matsumoto, M. Mino, Impact of the pattern layout on radio-frequency performance of thin-film SOI power MOSFETs, in *International Symposium on Power Semiconductor Devices and ICs*, Kitakyusha, Japan, 2004
68. A. Mai, H. Ruecker, R. Sorge, D. Schmidt, C. Wipf, Cost-effective integration of RF-LDMOS transistors in 0.13 μm CMOS technology, in *Silicon Monolithic Integrated Circuits in RF Systems*, San Diego, USA, 2009
69. R. Sorge, A. Fischer, A. Mai, P. Schley, J. Schmidt, C. Wipf, T. Mausolf, R. Pliquett, R. Barth, K.E. Ehwald, Complementary RF LDMOS module for 12 V DC/DC converter and 6 GHz power applications, in *Silicon Monolithic Integrated Circuits in RF Systems*, Phoenix, USA, 2011
70. D. Muller, A. Giry, D. Pache, J. Mourier, B. Szelag, A. Monroy, Architecture optimization of an N-channel LDMOS device dedicated to RF-power applications, in *International Symposium on Power Semiconductor Devices and ICs*, Santa Barbara, USA, 2005
71. B. Szelag, D. Muller, J. Mourier, F. Judong, A. Giry, D. Pache, A. Monroy, M. Roche, NLDMOS RF optimzation guidelines for wireless power amplifier applications, in *Bipolar/BiCMOS Circuits and Technology Meeting*, Santa Barbara, USA, 2005
72. B. Szelag, H. Baudry, D. Muller, A. Giry, D. Lenoble, B. Reynard, D. Pache, A. Monroy, Integration and optimization of a high performance RF lateral DMOS in an advanced BiCMOS technology, in *European Solid-State Device Research Conference*, Estoril, Portugal, 2003
73. D. Muller, A. Giry, F. Judong, C. Rossato, F. Blanchet, B. Szelag, A. Monroy, R. Sommet, D. Pache, O. Noblanc, High-performance 15-V Novel LDMOS transistor architecture in a 0.25-μm BiCMOS process for RF-power applications. IEEE Trans. Electron Devices **54**(4), 861–868 (2007)
74. K.E. Ehwald, B. Heinemann, W. Roepke, W. Winkler, H. Ruecker, F. Fuernhammer, D. Knoll, R. Barth, B. Hunger, H.E. Wulf, R. Pazirandeh, N. Ilkov, High performance RF LDMOS transistors with 5 nm gate oxide in a 0.25 μm SiGe:C BiCMOS technology, in *International Electron Devices Meeting*, Washington, USA, 2001

75. O. Bengtsson, A. Litwin, J. Olsson, Small-signal and power evaluation of novel BiCMOS-compatible short-channel LDMOS technology. Trans. Microw. Theory Techn. **51**(3), 1052–1056 (2003)
76. J.-P. Colinge, Fully-depleted SOI CMOS for analog applications. IEEE Trans. Electron Devices **45**(5), 1010–1016 (1998)
77. G.K. Celler, S. Cristoloveanu, Frontiers of silicon-on-insulator. J. Appl. Phys. **93**(9), 4955–4978 (2003)
78. S. Merchant, E. Arnold, H. Baumgart, S. Mukherjee, H. Pein, R. Pinker, Realization of high breakdown voltage (⩾700 V) in thin SOI devices, in *International Symposium on Power Semiconductor Devices and ICs*, Baltimore, USA, 1991
79. S. Merchant, E. Arnold, H. Baumgart, R. Egloff, T. Letavic, S. Mukherjee, H. Pein, Dependence of breakdown voltage on drift length and buried oxide thickness in SOI RESURF LDMOS transistors, in *International Symposium on Power Semiconductor Devices and ICs*, Monterey, USA, 1993
80. I.-J. Kim, S. Matsumoto, T. Sakai, T. Yachi, Analytical approach to breakdown voltages in thin-film SOI power MOSFETs. Solid-State Electron. **39**(1), 95–100 (1996)

Chapter 6
Lateral Power Transistors with Charge Compensation Patterns

Among the promising candidates for future LDMOS devices is the charge compensated LDMOS transistor which is focused on in this chapter. Charge compensation patterns have been successfully introduced in vertical superjunction MOSFETs [1]. A transfer of this topology to lateral power MOSFETs appears intriguing due to the reduction of drift resistance further beyond the one-dimensional silicon limit. Using a unit cell for lateral power MOSFETs, the charge compensation patterns are presented and their operation principle is explained. Then, different device designs for LDMOS transistors employing these charge-compensated drift regions are presented. The evaluation of electrical properties for these charge compensated LDMOS transistors yields low static power losses and reduced switching losses. The feasibility of integration of charge compensation patterns into smart-power ICs is evaluated considering the process technology required for formation of these patterns.

6.1 Concept of Charge Compensation Patterns for Superjunction Devices

In order to overcome the limitation of drift region resistance in lateral power transistors with a uniform doping concentration, the RESURF design has been successfully introduced as discussed in Chap. 5. The RESURF is based on the full depletion of the charge carriers in the n-doped drift region.

A further reduction of the drift region resistance in lateral power transistors has been previously demonstrated by the application of charge compensation patterns which are also referred to as superjunction or 3D-RESURF [2]. The fundamental unit cell design for a lateral charge compensation pattern which is suitable for integration into LDMOS transistors is depicted in Fig. 6.1.

Using a lateral charge compensation pattern, the conventional n-doped drift region is intersected by p-doped regions. Due to the similarity to vertical superjunction transistors, the active regions of the charge compensation patterns are also referred to as p- and n-pillars. By the incorporation of these pillars, the charges in

T. Erlbacher, *Lateral Power Transistors in Integrated Circuits*,
Power Systems, DOI 10.1007/978-3-319-00500-3_6

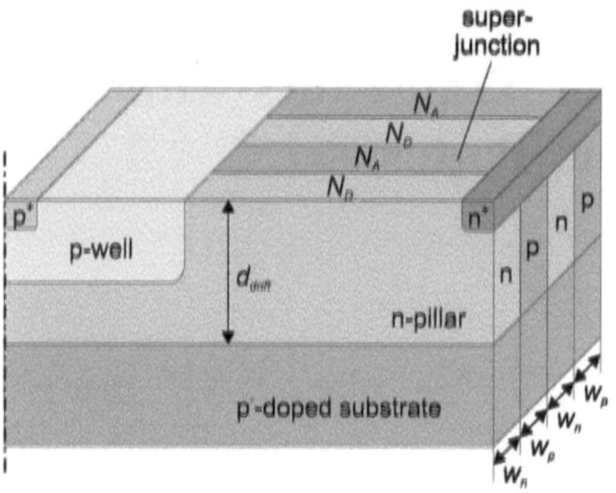

Fig. 6.1 Unit cell of a charge compensation pattern for lateral power devices (for simplicity only the semiconductor region is shown)

the drift region are not only compensated by the charges below the drift region as in a conventional RESURF design, but mainly laterally between the neighboring p- and n-pillars. This allows for the incorporation of a higher doping concentration in the n-drift region and further reduces the drift region resistance. This concept relies on the compensation of all charges in the n-drift pillars by the charges in the adjacent p-pillars at a blocking voltage well below the breakdown voltage. The voltage where the pillars are fully depleted is referred to as *pinch-off voltage*. Subsequently, a space charge region is formed across the full length of the drift region before a critical electric field develops from drain to source as shown in Fig. 6.2.

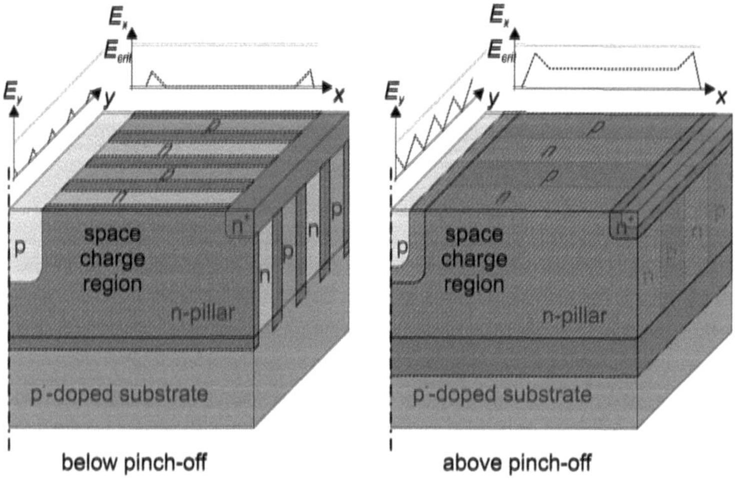

Fig. 6.2 Extension of space charge regions and electric field distribution in a unit cell of a charge compensation pattern below (*left side*) and above (*right side*) pinch-off voltage

The electric field distributions for a sound charge compensation pattern indicates that the electric field between the neighboring p- and n-pillars increases and decreases linearly (for a constant doping concentration) because the compensation takes place at low blocking voltages. As the pinch-off voltage is reached, all the mobile charges in the space charge region have been removed and a continuous space charge region has developed. A further increase of the drain-source voltage will result in the increase of the space charge region into the p^+ and n^+-regions at the boundaries to the charge compensation patterns. Since no more corresponding charges are available within the drift region, the electric field lines are not affected in the drift region due to the lack of counter-charges and a nearly constant electric field between drain and source develops. In a uniformly doped drift region, in contrast, an electric field in excess of the critical value would have developed for such a high doping concentration resulting in avalanche breakdown. Using a charge compensation pattern, the electric field between drain and source will remain below its critical value despite a high doping concentration in the drift region for voltages well above the breakdown voltage of a plane-parallel pn-junction. Note that the vertical electric field is kept below a critical value by using a lightly doped p-doped substrate or n-buffer layer—depending on the device design [3]. More sophisticated designs even rely on a dielectric layer (silicon-on-insulator material) or an insulating semiconductor like sapphire for electric insulation in vertical direction [2, 4].

The charge compensation patterns only act on the drift region resistance making this approach particularly feasible for devices designed for operation voltages beyond 50 V. The benefit of these superjunction cells on the on-state resistance is discussed next. Particularly, the impact of breakdown voltage is reviewed.

Regarding the benefits of charge compensation patterns in power semiconductor devices, the minimum area-specific drift region resistance of vertical superjunction MOSFETs, which has been previously derived [5], should be considered first for further discussion:

$$\frac{R_{drift,SJ=VDMOS} \cdot A}{\Omega \mathrm{cm}^2} = 1.98 \times 10^{-1} \cdot (d/\mathrm{cm})^{5/4} \cdot \left(\frac{BV}{\mathrm{V}}\right). \tag{6.1}$$

In vertical devices, the dependence of the drift region resistance on breakdown voltage is less pronounced than for homogeneously doped drift regions or compared to lateral power transistors with RESURF design. This incorporation of vertical charge compensation patterns allows for a significant reduction of drift region resistance compared to a homogeneously doped drift region in vertical power transistors for 600 and 900 V applications.

Following a similar approach, the optimum drift region resistance of lateral charge compensation patterns can be estimated analytically for a design with equidistant p- and n-pillars of width W_n and W_p (equal to half-cell width W_{SJ}) of identical doping concentrations N_{drift} and N_A, respectively. The resistance of the charge compensation patterns in the drift region shown in Fig. 6.1 can be expressed as

$$R_{drift,CCP} = \frac{2}{q\mu_{n,drift} N_{drift,CCP}} \cdot \frac{L_{drift}}{d_{drift} \cdot (W_n + W_p)}, \tag{6.2}$$

which is twice as high as in the case of a (continuous) RESURF-based drift region given by (5.4). This is due to the p-pillars in the charge compensation patterns which cannot conduct the electron current in on-state condition. The impact of the space-charge region at the junction of adjacent p- and n-pillars on the drift region resistance in (6.2) has been neglected. This is valid as long as the high doping concentrations of the p- and n-pillars at low voltages in forward conduction results in a small space-charge region compared to the width of the pillars. Taking the area of the drift region into account, the *area-specific drift region resistance* is

$$R_{drift,CCP} \cdot A = \frac{2}{q\mu_{n,drift} N_{drift,CCP}} \cdot \frac{L_{drift}^2}{d_{drift}}. \tag{6.3}$$

The vector of the electric field inside these charge compensation patterns consists of a component perpendicular to the p- and n-pillars E_x—where the electric field increases and decreases—and a component parallel to these pillars E_y—where a nearly constant electric field occurs between drain and source. Neglecting the vertical electric field, the absolute value of the critical electric field E_{crit} across the charge compensation pattern at the onset of avalanche breakdown is determined by

$$E_{crit} = \sqrt{E_{x,max}^2 + E_{y,bd}^2}. \tag{6.4}$$

Avalanche breakdown is triggered by the increase of the parallel electric field component $E_{y,bd}$ at a corresponding perpendicular electric field $E_{x,max}$ at the pn-junctions of the charge compensation patterns. The orthogonal electric field component is governed by the formula for a pn-junction with a maximum value of

$$E_{x,max} = \frac{qN_{opt} W_{SJ}}{2\varepsilon_0 \varepsilon_r} = \alpha E_{crit}. \tag{6.5}$$

Introducing the scalar α as the portion of the electric field perpendicular to the pillars as defined in (6.5) leads to the electric field parallel to the pn-junctions at breakdown

$$E_{y,bd} = \sqrt{1 - \alpha^2} E_{crit}. \tag{6.6}$$

For a feasible charge compensation design, the breakdown voltage can be approximates as

$$BV \approx E_{y,bd} L = \sqrt{1 - \alpha^2} E_{crit} L. \tag{6.7}$$

Combination of the previously derived equations yields an area-specific drift region resistance of

$$\begin{aligned} R_{drift,CCP} \cdot A &= \frac{1}{\mu_{n,drift}\varepsilon_0\varepsilon_{Si}\alpha E_{crit}} \cdot \frac{L_{drift}^2 \cdot W_{SJ}}{d_{drift}} \\ &= \frac{1}{\mu_{n,drift}\varepsilon_0\varepsilon_r\alpha(1-\alpha^2)E_{crit}^3} \cdot \frac{W_{SJ}}{d_{drift}} \cdot BV^2 \end{aligned} . \tag{6.8}$$

For an optimized charge compensation pattern which is much longer than deep, (6.8) becomes minimal at

$$\alpha = \frac{E_{x,max}}{E_{crit}} = \frac{1}{3}\sqrt{3} \approx 0.577. \tag{6.9}$$

Using Fulop's approximation for the avalanche breakdown in silicon, the following relationship between the critical electric field and the breakdown voltage can be established for the optimum doping concentration (6.5):

$$\frac{E_{crit,Si}}{\mathrm{Vcm}^{-1}} = 7.95 \cdot 10^5 \cdot \left(\frac{BV}{\mathrm{V}}\right)^{-\frac{1}{6}}. \tag{6.10}$$

For silicon, the area-specific drift region resistance of a charge compensation pattern can thus be approximated to be

$$\frac{R_{drift,CCP,Si} \cdot A}{\Omega\mathrm{cm}^2} = 4.52 \cdot 10^{-9} \cdot \frac{W_{SJ}}{d_{drift}} \cdot \left(\frac{BV}{\mathrm{V}}\right)^{2.5}. \tag{6.11}$$

Note that an exact analytical calculation of the optimum drift region resistance of the charge compensation patterns is neither feasible nor necessary for the discussion in the scope of this textbook due to depletion effects and current spreading from the channel into the drift region. Instead, this drift region resistance should be understood as a measure to compare this concept to state-of-the-art solutions. In Fig. 6.3, the further reduction of drift region resistance compared to the homogeneous doping discussed in Chap. 4 using (4.16) and (4.25) and the RESURF designs introduced in Chap. 5 using (5.10) are depicted.

The superjunction design does not yield an intrinsic advantage over the RESURF design in lateral power transistors. In fact, the dependence of the drift region resistance on the breakdown voltage to the power of 2.5 is similar for both designs. In contrast to the RESURF design, the drift region resistance of the charge compensation patterns can be affected by additional geometric parameters, namely the depth and the width of the charge compensation regions. For drift region resistance

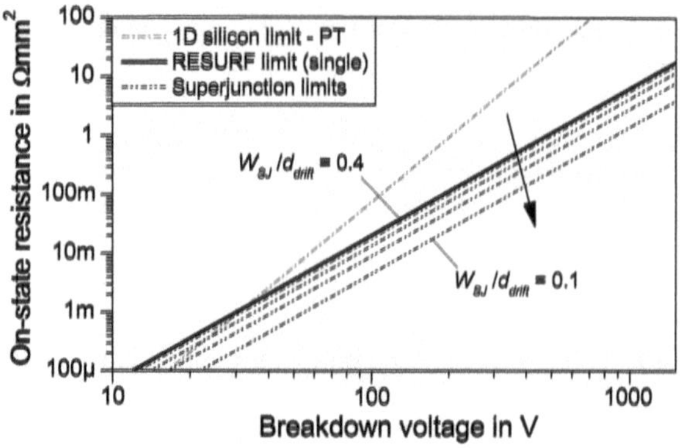

Fig. 6.3 Theoretical limits of drift region resistances for different breakdown voltages using homogeneous doping (1D limit) with 1 μm depth, RESURF design and charge compensation patterns with width to depth ratios ranging from 0.4 down to 0.1

of superjunction patterns to become significantly lower compared to the RESURF design, a ratio between the width of the pillars and their depths of less than 0.2 is desirable, i.e. the pillars should be 5 times deeper than wide. For very densely packed pillar structures, a high doping concentration can be used. An empirical relation between the optimum doping concentration and the pillar width has been deduced for vertical superjunction MOSFETs.

$$\frac{N_{opt,theo}}{\mathrm{cm}^{-3}} = 1.2 \cdot 10^{12} \left(\frac{W_{SJ}}{\mathrm{cm}}\right)^{-\frac{8}{7}}. \tag{6.12}$$

This relationship is also valid for lateral charge compensation patterns. It implies that the drift region resistance achieved by charge compensation patterns depends on the capability and limitations of the available processing technology.

6.2 Processing Technology for Charge Compensation Patterns

As indicated by (6.5), the doping concentration and width of the pillars are correlated to physical properties of the semiconductor material, i.e. the dielectric constant and the critical electric field (which depends on the impact ionisation rate). The relationship between the width of the n-doped and p-doped regions and the drift region resistance in silicon technology is shown in Fig. 6.4.

Similar to a RESURF design, fabrication of charge compensation patterns requires precise control of the doping concentrations. On a production scale, this precision can be achieved by means of ion implantation. In contrast to the RESURF

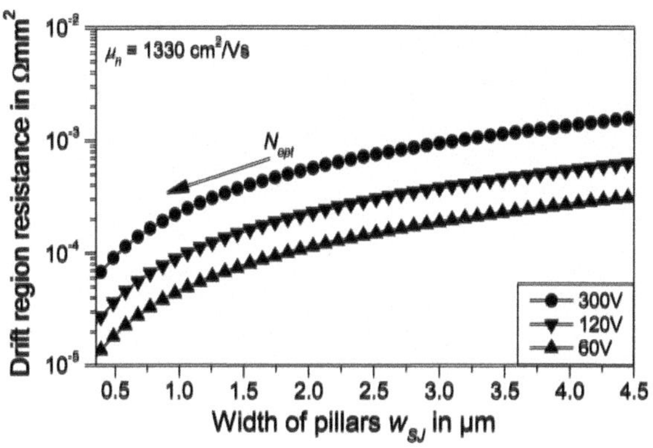

Fig. 6.4 Drift region resistance for different widths of equidistant lateral charge compensation patterns in silicon without interdiffusion for different breakdown voltages

design—compare (5.3)—there is no universal optimum implantation dose for achieving charge balance in superjunction LDMOS transistors. Instead, the width of the charge compensation patterns needs to be carefully considered. This includes effects like lateral struggle, diffusion and overlay accuracy for the photo lithography. An unintentional charge imbalance, i.e. a deviation between the charges present in the p- and n-regions, will cause a significant reduction in breakdown voltage of the charge compensation patterns [6, 7].

Generally, two limitations of the fabrication of charge compensation patterns using ion implantation should be considered.

- Firstly, the depth of wells created by ion implantation is limited by the implantation energy. While this is an even more severe constraint for the fabrication of vertical superjunction devices, the requirements regarding the depth of the wells are also high for lateral superjunction devices. As shown in the previous section, deep charge compensation patterns allow for a lower drift region resistance. But it should also be noted that the reduction of drift region resistance for deeper patterns will be diminished by the increasing current spreading from the planar channel (at the gate) into the drift region and, from there, to the drain electrode.
- Secondly, the activation of dopants and the fabrication of deep wells by diffusion at elevated temperatures (e.g. 900 °C and above in silicon) results in interdiffusion of dopants between the charge compensation regions, effectively reducing the local doping concentration. Since p- and n-dopants exhibit different diffusion rates, a local charge imbalance is created at the pn-junctions of the charge compensation regions. This effect was previously studied for vertical superjunction devices in detail [8]. The findings can be carried over to superjunction patterns for lateral power devices. The interdiffusion of boron and phosphorous in charge compensation patterns is depicted in Fig. 6.5 for various annealing conditions using the complementary error function for the diffusion process [9].

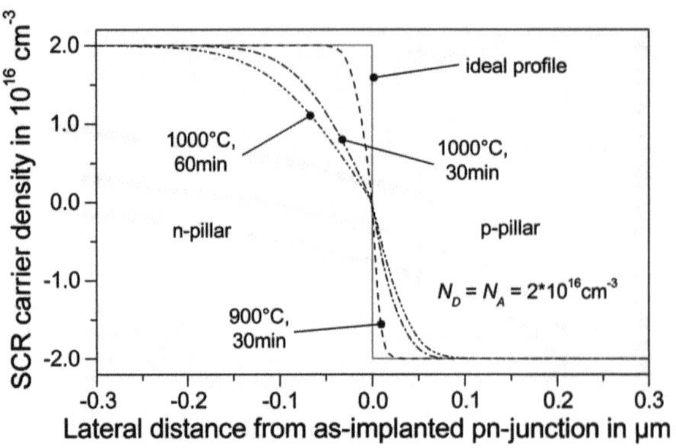

Fig. 6.5 Doping profiles for equidistant lateral charge compensation patterns in silicon for several annealing conditions; boron and phosphorous are considered as acceptor and donor atoms

Here, the lateral struggle of the ion implantation was neglected. For a given thermal budget, the optimum doping concentration in charge compensation patterns which takes diffusion into account has been extracted for a given fabrication process (including gate oxide growth, p-body drive-in and source/drain annealing) [8] as shown in Fig. 6.6.

The approximations in the underlying model result in some limitations towards very small pillar widths. Nevertheless, it is illustrated that the interdiffusion becomes particularly critical for small widths of the charge compensation regions across all desired blocking voltages and when the fabricated charge compensation patterns are subject to high thermal budgets. It is also evident, that lateral superjunction devices in silicon could be implemented with pillar widths in the submicron range allowing for significantly higher doping concentrations than a

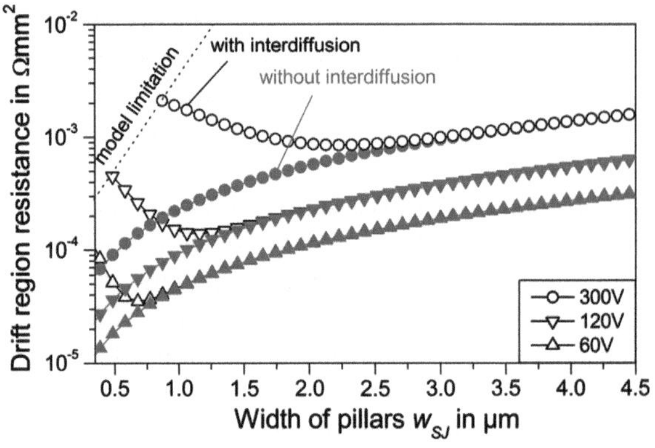

Fig. 6.6 Drift region resistance for different widths of equidistant lateral charge compensation patterns in silicon without and with interdiffusion for a given fabrication process [8]

RESURF-design facilitates. However, the thermal budget of the chosen fabrication technology has to be considered. Additionally, neglecting the space charge region width in the pillars in forward conduction is not valid for very narrow pillars in the sub-micron range. TCAD simulations using ideal pillar shapes indicate that the extension of space-charge regions at the built-in potential becomes significant for pillar widths of approximately 600 nm [6]. In fact, a reduction of pillar width below 350 nm results in an increase of drift region resistance even for higher doping concentrations.

6.3 Device Designs with Charge Compensation

6.3.1 Fundamental Device Design

The fundamental layout of a superjunction LDMOS transistor with symmetric charge compensation patterns is presented in Fig. 6.7.

It is based on the equidistant charge compensation patterns described in Fig. 6.1. The key benefit is a further reduction of drift region resistance compared to RESURF-based LDMOS transistors for significantly deep pillars. From this fundamental design a wide range of prospective superjunction LDMOS transistors has been proposed. To foster the understanding of prospects and limitations of charge compensation patterns in LDMOS transistors, these design considerations are reviewed.

Another benefit of the SJ LDMOS transistor is the reduction of the electric field at the surface of the drift region [10, 11]. In the SJ LDMOS transistor, the electric field peak can be situated in the bulk near the drain electrode, making the transistor more reliable, especially with respect to avalanche ruggedness and hot carrier injection effects compared to conventional designs.

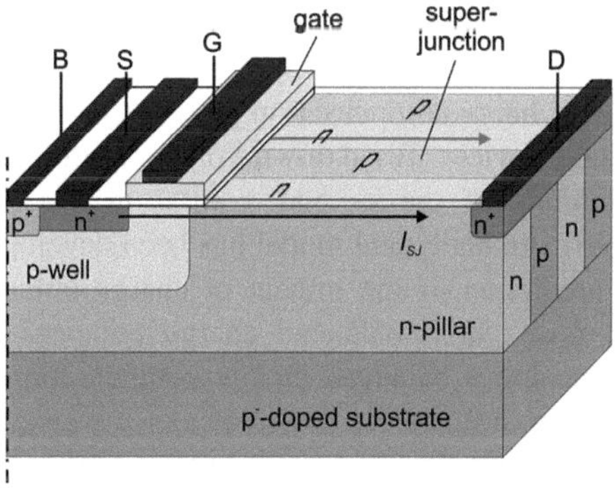

Fig. 6.7 Superjunction LDMOS transistor with balanced charge compensation patterns

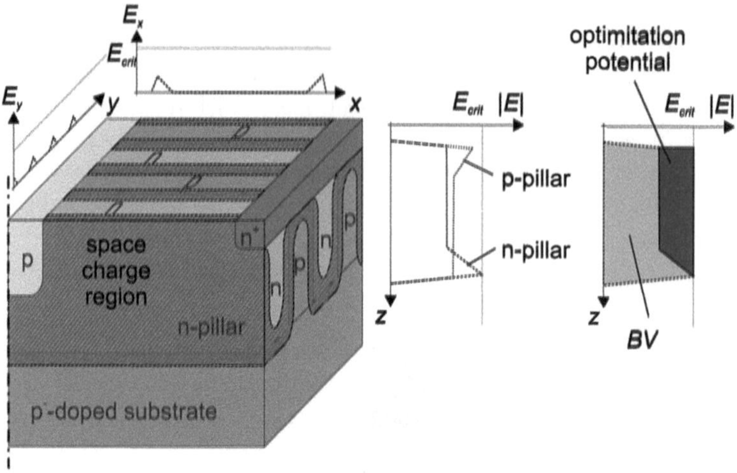

Fig. 6.8 Substrate depletion effect in superjunction LDMOS transistors: charge imbalance caused by substrate acceptor charges; electric field distribution along pillar depth; reduction of breakdown voltage due to charge imbalance

6.3.2 Substrate-Assisted Depletion Effects

A major limitation of the design shown in Fig. 6.7 is the unintentional charge imbalance that is introduced by the p-doped substrate. This so-called substrate-assisted depletion effect [12] is depicted in Fig. 6.8.

Due to the presence of the acceptor atoms of the p-doped substrate, additional negative charges are present in the charge compensation patterns which compensate a portion of the positively charged donor atoms. Thereby, charge imbalance in the charge compensation patterns occurs, and a vertical electric field is generated. Moreover, the highly doped drain contact and the p-body of the gate region in the superjunction LDMOS transistor also create imbalance by supplying additional donor and acceptor charges, respectively. Hence, care must be taken to minimize charge imbalance when designing the charge compensation patterns and the LDMOS device.

Charge imbalance in charge compensation patterns has been extensively studied in vertical superjunction devices. Breakdown voltage is strongly deteriorated when the doping concentrations of the p- and n-pillar in the charge compensation structure mismatch [13]. An analytical model has been developed by Napoli, Wang and Udrea [14] for prediction of the impact of charge imbalance on breakdown voltage. For this purpose, an imbalanced charge compensation pattern can be modelled by superimposing a balanced charge compensation pattern with a pin-diode as illustrated in Fig. 6.9.

An imbalanced equidistant charge compensation pattern is assumed with pillar width W_{SJ} and doping concentrations N_A and N_D for the p- and n-doped pillars,

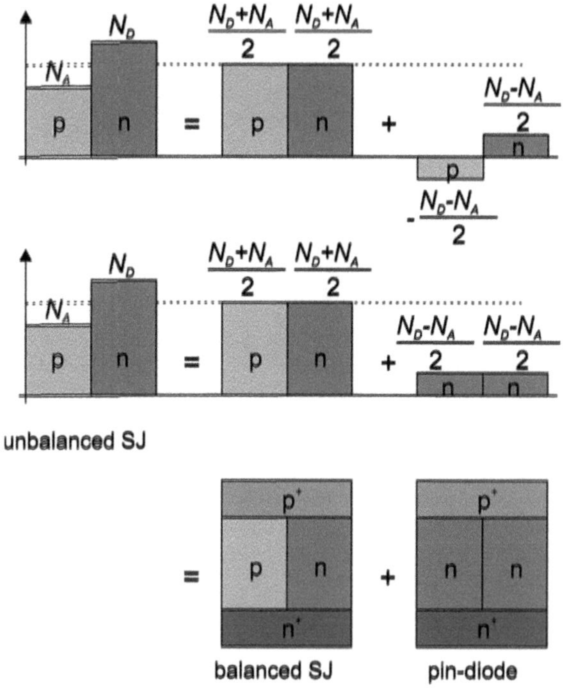

Fig. 6.9 Modelling of charge imbalance by superimposing a balanced superjunction with a pin-diode, adopted from [14]

respectively. Additionally, the charge compensation pattern is operated in punch through condition. Then, the electric field distribution in this structure is found to be equal to the superimposed electric fields that are present in a *balanced charge compensation pattern* with doping concentration

$$N_{SJ} = \frac{N_D + N_A}{2}, \tag{6.13}$$

and a pin-diode with a homogeneous doping concentration in its drift region of

$$N_{pin} = \left|\frac{N_D - N_A}{2}\right|, \tag{6.14}$$

respectively. A more complex model has also been derived which is also valid in the case of non-punch through conditions [15]. For practical use in LDMOS devices, however, a design with a breakdown voltage which corresponds to a non-punch through condition is not feasible with respect to minimum cell geometry.

In order to judge the mismatch between doping concentrations in adjacent pillars, a value for the charge imbalance *CI* is introduced, which is given by

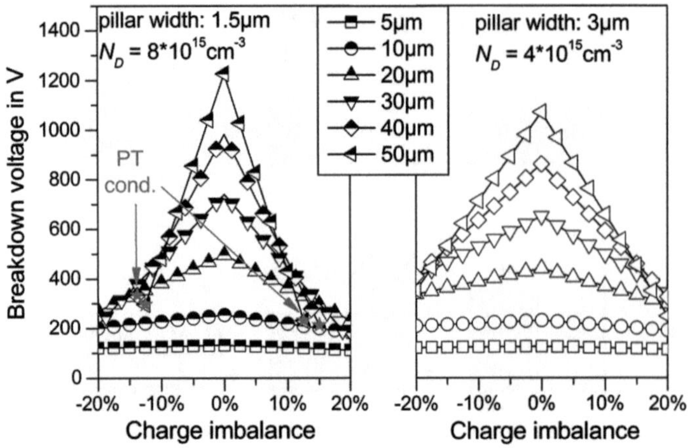

Fig. 6.10 Deterioration of breakdown voltage due to charge imbalance between n- and p-pillars in lateral charge compensation patterns for different drift lengths and p-pillar doping concentrations; transition from punch through (PT) to non-punch through (NPT) condition is marked by *grey stars*

$$CI = \frac{(N_A - N_D)}{N_{nominal}}. \tag{6.15}$$

A positive value of *CI* represents a surplus of negative space charges from an excessive p-doping [13]. A deficiency of positive space charges from insufficient n-doping is equivalently indicated by a negative *CI* value. Using the above analytical model, the breakdown voltage for charge compensation pattern with various degrees of charge imbalance can be deduced from Fig. 6.10.

The degradation of breakdown voltage for a given charge imbalance is intensified when narrower pillars with higher doping concentration are used. The substrate-assisted depletion effect causes similar deterioration in superjunction LDMOS transistors for a balanced charge compensation pattern on a p-doped substrate. The breakdown voltage in investigated superjunction LDMOS transistors with a balanced charge compensation pattern directly on a p-doped substrate is well below theoretical limits [3].

6.3.3 Charge-Balancing Technologies in Junction Isolated Devices

Based on the origin of the charge imbalance, several technologies have been published to overcome substrate-assisted depletion effects in SJ LDMOS transistors.

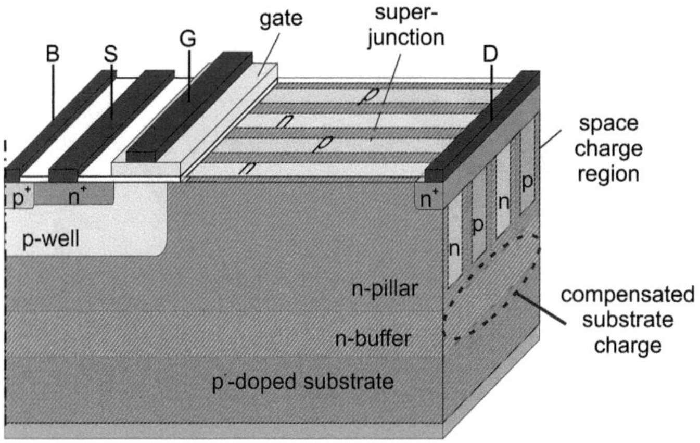

Fig. 6.11 Implementation concept of n-buffer with positive charge surplus to compensate substrate-assisted depletion in lateral charge compensation patterns, as introduced by [7]

6.3.3.1 Buffer Layer Between Charge Compensation Patterns and Substrate

An introduction of a homogeneously doped n-buffer layer has been proposed to compensate the electric field originating from the p-doped substrate [7, 16]. The idea is to provide an additional vertical compensation structure to shield the substrate from the charge compensation patterns by providing an adequate number of positive donor charges. A feasible topology using an n-doped buffer layer is presented in Fig. 6.11.

In forward conduction, the n-buffer layer is also capable of conducting electron current and can result in a further reduction of the drift region resistance. However, this effect is not significant due to the low doping concentration in comparison to the n-pillars in the charge compensation patterns. Still, LDMOS transistors with total device resistance of 266 mΩ mm^2 and a breakdown voltage of 87.5 V were experimentally demonstrated using this topology [7].

Given that epitaxial processes in silicon are readily available and that ion implantation technology could also be used, the incorporation of an n-buffer layer for compensation of the charges from the p-doped substrate requires little additional fabrication effort beyond the fabrication of the charge compensation patterns themselves. The design is also robust to up to ±5 % of charge imbalance, providing a decent process window. Moreover, the n-buffer layer enables high-side operation of this device because the drain-to-substrate voltage can be blocked across the p-substrate/n-buffer junction when a proper device design is used.

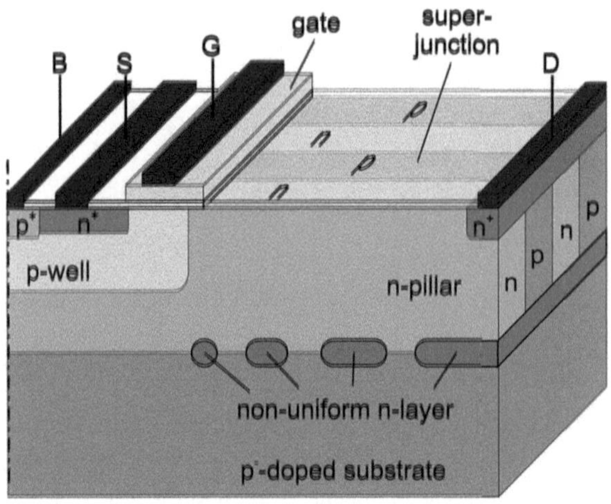

Fig. 6.12 Implementation concept of a non-uniformly doped buffer layer to compensate substrate-assisted depletion in lateral charge compensation patterns, as described by [17]

6.3.3.2 Non-uniformly Doped Buried Layer

Instead of a homogeneously doped buffer layer, a non-uniformly doped buried layer as depicted in Fig. 6.12 can also be used to promote adequate charge compensation [17].

The spacing of the n-doped islands composing the buried layer is designed to achieve compensation of the substrate-assisted depletion effect near the drain electrode. In this manner, the electric field at the interface between the p-doped substrate and the pillars is adjusted. Instead of a purely vertical component, the vector of the electric field is shifted towards the drain region. Therefore, the breakdown voltage will be higher when the critical electric field is reached. From numerical simulation, total drift region resistances of 4.4 Ω mm^2 and 19 Ω mm^2 were estimated for LDMOS transistors with breakdown voltages of 774 and 1,304 V, respectively.

6.3.3.3 Imbalanced Charge Compensation

Similarly, the surplus of negative acceptor charges arising from the substrate can also be compensated by control of the width of the p-doped and n-doped pillars in the charge compensation patterns [11]. Providing an excess of positive donor charges can be implemented to restore charge balance. Compared to a balanced charge compensation pattern, this intentional imbalance patterns can be achieved by any of the following design modifications:

- Increasing the width of the n-regions
- Reducing the width of the p-regions
- A higher doping concentration of the n-regions
- A lower the doping concentration of the p-regions.

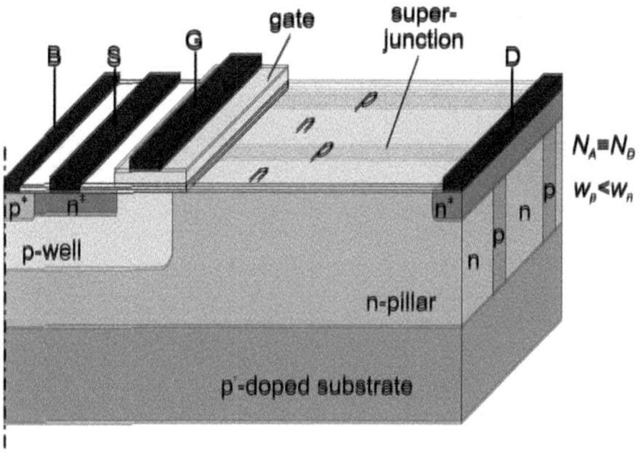

Fig. 6.13 Implementation concept of p-regions with negative charge deficiency to compensate substrate-assisted depletion in lateral charge compensation patterns

Note that these approaches may require further modifications of the pillar topology to ensure pinch-off and high breakdown voltage. An example for an intentionally imbalanced charge compensation pattern employing smaller p-regions is provided in Fig. 6.13.

Fabricated devices with this adjustment to pillar widths yielded a reported total device resistance of 3.53 Ω mm^2 at a breakdown voltage of 335 V. A similar methodology can be applied to compensate the charge imbalance created by the p-body and drain regions.

6.3.3.4 Combination of Buffer Layer and Imbalanced Charge Compensation

By combining the homogeneously doped n-buffer layer and imbalanced charge compensation patterns, breakdown voltages close to the theoretical limit can be achieved [3]. Using TCAD simulations, a suitable charge compensation design was derived which takes vertical and lateral electric fields into consideration. This integration concept is depicted in Fig. 6.14.

This concept also offers straight-forward integration capability into Smart-Power ICs. The shape of the p- and n-pillars can be determined by device geometry. Hence, tapering the shape is a feasible method to increase breakdown voltage.

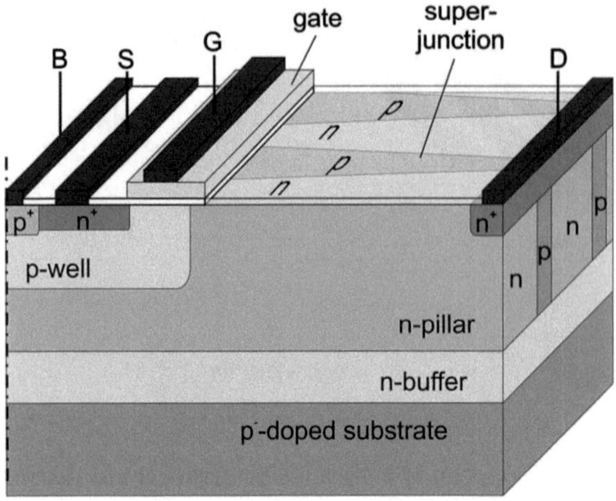

Fig. 6.14 Implementation concept of n-buffer layer and imbalanced charge distribution to overcome substrate-assisted depletion in lateral charge compensation patterns, adopted from [3]

6.3.3.5 Combination of Charge Compensation Patterns with RESURF Topology

Another approach towards the suppression of substrate-assisted depletion effects is the combination of superjunction and RESURF structures [6]. The corresponding device topology is indicated in Fig. 6.15.

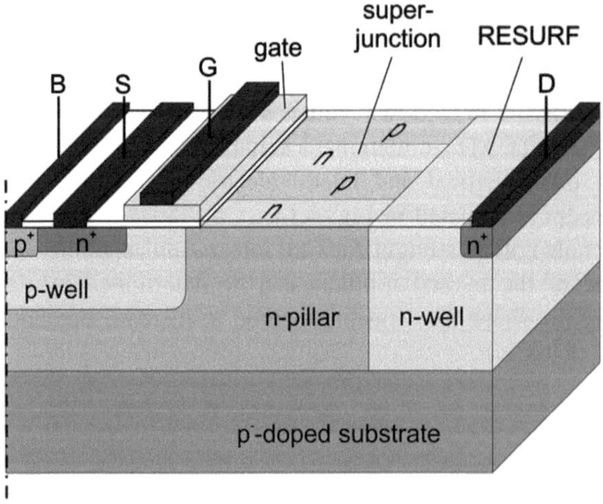

Fig. 6.15 Implementation concept of a combination of superjunction and RESURF topologies to overcome substrate-assisted depletion in lateral charge compensation patterns, as described in [6]

In this design, the charge compensation patterns are terminated by a narrow RESURF region near the drain contact. The RESURF region exhibits a distinct length necessary to compensate the substrate-assisted depletion effect. If the RESURF region is too short, this compensation is incomplete and the breakdown voltage is deteriorated. Whereas, a long RESURF region results in an increased drift region resistance. By proper selection of design parameters, total device resistances of 0.342 and 4.88 Ω mm^2 have been calculated for LDMOS transistors with breakdown voltages of 121 and 718 V, respectively.

Similar considerations have led to the approach of integrating a charge compensation pattern in the top n-drift region of a multi-acting RESURF design as described in Fig. 6.16.

This device design exploits the validity of the RESURF design even if the part of the drift region is replaced by an adequately doped charge compensation pattern. Thereby, a higher doped drift region with a larger cross-section becomes available for current conduction. This design using a state-of-the-art double-acting RESURF design offers an on-state device resistance of 54 mΩ mm^2 for an n-channel LDMOS transistor with a breakdown voltage of 61.1 V. TCAD simulations indicate that the total device resistance can be further reduced to 36.2 mΩ mm^2 for a device with 60 V breakdown voltage when the charge compensation pattern is incorporated into a multi-acting RESURF [18]. Of course, this also requires additional fabrication effort.

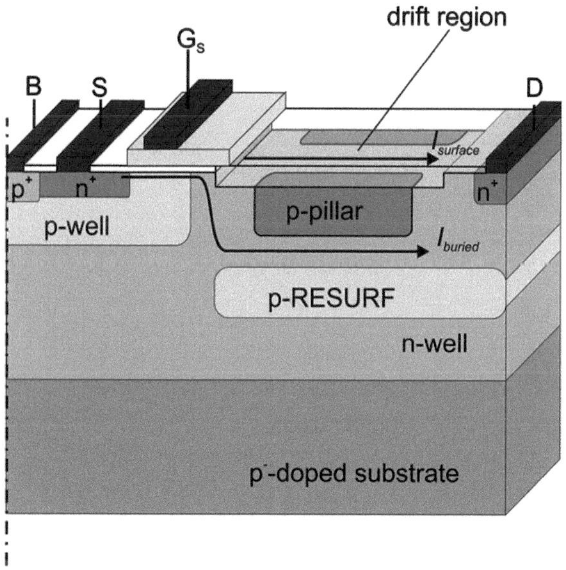

Fig. 6.16 Combination of superjunction and RESURF topologies to minimize drift region resistance, adopted from [18]

6.3.4 Charge-Balancing Technologies in Dielectric-Isolated Devices

Even though the reduction of drift region resistance is particularly powerful for high voltage devices, superjunction technology has also been investigated in LDMOS transistors for fast switching applications on silicon-on-insulator substrates. The fundamental device geometry for this technology as depicted in Fig. 6.17 has been previously proposed [19].

Since the electric field between drain and source also develops across the buried oxide, substrate charges will be influenced. In particular, it has been shown in Chap. 5 that the buried layer can be treated as a lightly doped substrate even enabling a RESURF effect when the top silicon layer is thin enough [20]. Therefore, charge compensation patterns on SOI are subject to substrate-assisted depletion effects similar to those fabricated using junction-isolation. Equivalent considerations hold true, when the substrate is imagined as a field plate that acts across the buried oxide onto the charge compensation patterns.

A charge-imbalance technology for superjunction LDMOS transistors using SOI technology has been proposed [2]. The integration concept shown in Fig. 6.18 resembles the imbalanced charge compensation design introduced in Fig. 6.14.

In this design, the n-pillar on the insulator is tapered to increase towards the n^+-drain region. Thereby, the field effect of the substrate on the charge compensation patterns can be affected. By intentionally misbalancing the charge compensation patterns towards more positively charged donors in the n-pillar, negative charges are influences in the substrate. Here, the net charge balance equals to the positive donor charges of the n-pillar which is equal to the negative acceptor

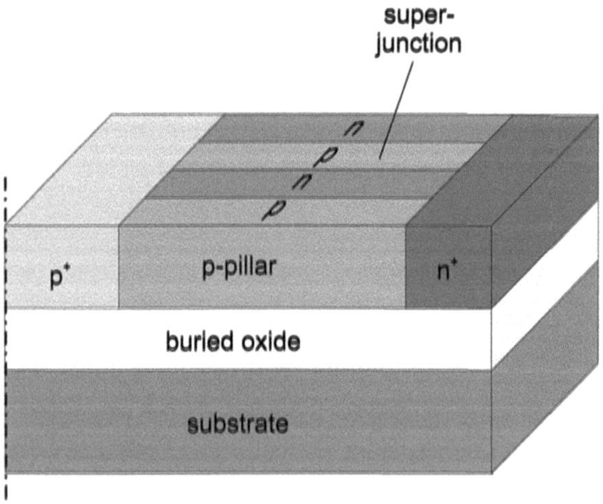

Fig. 6.17 Fundamental charge compensation patterns for superjunction LDMOS transistors in silicon-on-insulator technology, adopted from [19]

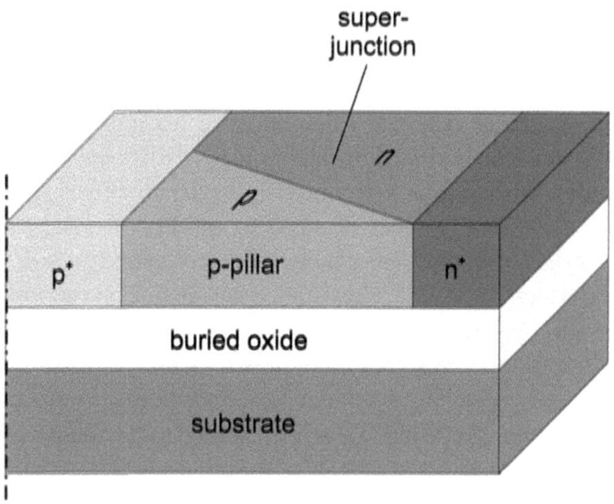

Fig. 6.18 Implementation concept of intentionally imbalanced charge distribution using silicon-on-insulator technology and net charge optimization [21]

charges in the smaller p-pillar plus the negative charges in the substrate. As this constitutes to a RESURF like behavior of the substrate, an almost constant lateral electric field and a high breakdown voltage close to the theoretical limit can be achieved [21].

6.3.5 Charge Compensation Patterns Beyond CMOS Technology

To conclude the incorporation of charge compensation patterns in lateral power transistors, technology developments beyond standard CMOS technology are briefly presented.

In order to prevent substrate depletion effects in SOI, the local removal of the substrate under the charge compensation patterns by wet or dry chemical etching is an effective method [22]. In addition, the drain-to-substrate capacitance across the insulating layer is completely removed. Without the presence of a substrate, charges cannot be influenced. The blocking voltage is therefore completely absorbed in the charge compensation patterns and the insulator. Unfortunately, the local removal of the substrate requires processing steps that are not routinely available in standard CMOS processes.

Instead of using an SOI substrate, the implementation of lateral power semiconductor devices on thick insulating substrates was proposed [23]. Given that the monocrystalline silicon layer must be epitaxially grown, the substrate also needs to exhibit a monocrystalline lattice structure. The material system of choice for this purpose is silicon-on-sapphire (SOS). Basically, the thick insulator provides the same benefits as the back etched SOI material regarding the implementation of charge compensation patterns [24, 25]. This means that a high constant electric field can be achieved using SOS-based superjunction LDMOS transistors, which exhibit low drift region resistance and low gate charge. Silicon-on-sapphire substrate can also be processed in standard CMOS technology. However, fabrication costs for silicon-on-sapphire substrates are significantly higher than for silicon-on-insulator substrates [22]. Additionally, heat dissipation through the sapphire substrate is hampered due to the lower thermal conductivity compared to silicon.

A significant reduction of device resistance has also been reported by combining charge compensation patterns and FinFET technology [24, 25]. These technologies target LDMOS transistors with voltage ratings of 100 V and below. As discussed, the incorporation of the charge compensation patterns minimizes drift region resistance. Additionally, the FinFETs achieve larger equivalent channel width per unit cell, and the channel resistance is also reduced. The concept of reducing channel resistance using trenches and FinFET technology will be reviewed in Chaps. 7 and 8 in detail. The fabrication process for the SJ-FinFET devices itself is compatible with advanced CMOS technology. However, device fabrication requires deep trench isolation and a gate trench formation which are not readily available in foundry-based processes because smart-power IC fabrication lacks behind state-of-the-art CMOS technology in terms of technology nodes, and FinFETs have just recently been introduced for fabrication of highly integrated CMOS circuits. Still, this technology makes up for a promising approach to reduce total device resistance when FinFET technology becomes available in smart-power IC fabrication.

6.3.6 Comparison of Superjunction LDMOS Transistor Designs

Breakdown voltage, on-state resistance and gate charge (if disclosed) for different superjunction LDMOS devices are displayed for fabricated devices in Table 6.1 and for TCAD simulations in Table 6.2.

6.4 Electrical Properties of Charge Compensated Lateral Power Transistors

As previously discussed, the advantages of superjunction LDMOS transistors are the reduction of total device resistance in on-state and the reduction of device capacitances. These benefits are discussed in this section.

Table 6.1 Breakdown voltage, on-state resistance and gate charge measured from fabricated LDMOS transistors using charge compensation patterns reported in literature

BV (V)	$R_{DS,on}$ (Ω mm^2)	Q_{gd} (nC/mm^2)	Device type	Substrate	Source
335	3.53	–	SJ-LDMOS	Bulk Si	[11]
87.5	0.266	–	SJ-LDMOS	Bulk Si	[7]
605	8.7	–	SJ-LDMOS	Bulk Si	[26]
718	21.56	–	SJ-LDMOS	Bulk Si	[10]
61.1	0.054	–	SJ-LDMOS w/2D-RESURF	Bulk Si	[18]
520	82	–	SJ-LDMOS	SOS	[23]
170	8.7	–	SJ-LDMOS	SOS	[27]
250	11.4	–	SJ-LDMOS	SOS	[28]
100	0.12	–	SJ-LDMOS	SOI	[24]
54	0.056	–	SJ-LDMOS	SOI	[24]
98	0.082	–	SJ-FinFET	SOI	[24]
51	0.035	–	SJ-FinFET	SOI	[24]

Table 6.2 Breakdown voltage, on-state resistance and gate charge from simulated LDMOS transistors using superjunction patterns reported in literature

BV (V)	$R_{DS,on}$ (Ω mm^2)	Q_{gd} (nC/mm^2)	Device type	Substrate	Source
703	11.52	–	SJ-LDMOS	Si	[10]
718	4.88	–	SJ-LDMOS w/RESURF	Si	[6]
121	0.342	–	SJ-LDMOS w/RESURF	Si	[6]
1,244	21.7	–	SJ-LDMOS w/RESURF	Si	[6]
774	4.4	–	SJ-LDMOS	Si	[17]
188	0.32	–	SJ-LDMOS	Si	[17]
1,304	19	–	SJ-LDMOS	Si	[17]
60	0.0362	–	SJ-LDMOS w/multi-RESURF	Si	[18]
1,330	26	–	SJ-LDMOS	SOS	[23]
630	6.1	–	SJ-LDMOS	SOI	[2]
160	0.9	–	SJ-LDMOS	SOI	[2]
87	0.223	–	SJ-LDMOS	SOI	[25]
317	4.83	–	SJ-LDMOS	Si membrane	[22]
300	0.86	–	SJ-LDMOS	Si membrane	[22]
154	0.165	1.39	SJ-LDMOS	SOI	[29]

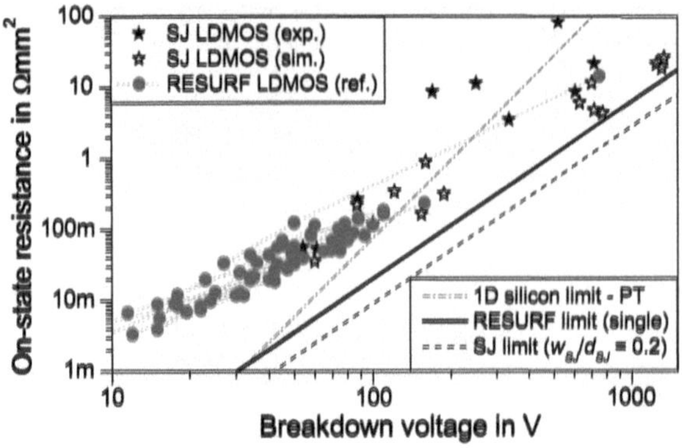

Fig. 6.19 Trade-off between breakdown voltage and on-state resistance in superjunction LDMOS transistors, data from Tables 6.1 and 6.2

6.4.1 Reduction of On-State Resistance

In Fig. 6.19, the comparison of superjunction LDMOS transistor designs regarding their on-state resistance performance using the trade-off with breakdown voltage is shown.

The total devices resistance of LDMOS transistors for integrated circuits approaches the theoretical superjunction limit for breakdown voltages beyond 100 V due to dominant contribution of drift region resistance, and the one-dimensional silicon limit is easily surpassed. Regarding the application for medium and low voltage devices below 100 V, the superjunction limit cannot be exploited due to the dominating contribution of channel resistance. This circumstance is further investigated in Chap. 7.

Figure 6.19 also indicates that the implementation of superjunction device topologies for LDMOS transistors do not generally represent significant progress beyond state-of-the-art double-acting RESURF transistors with respect to total device resistance. Simulation results predict lower total device resistances especially for high voltage devices, but electrical characterization of fabricated superjunction LDMOS transistors falls short of these expectations. Reasons can be found in both the charge imbalance and the diminishing returns from deep superjunction patterns with low drift region resistance due to limited current spreading from the channel region. Nevertheless, the incorporation of charge compensation patterns yields other benefits making this topology feasible to implement.

6.4.2 Impact on Device Capacitances

Switching losses in LDMOS transistors for energy conversion can be reduced by minimizing the output capacitance. At low blocking voltages, the pillars are not

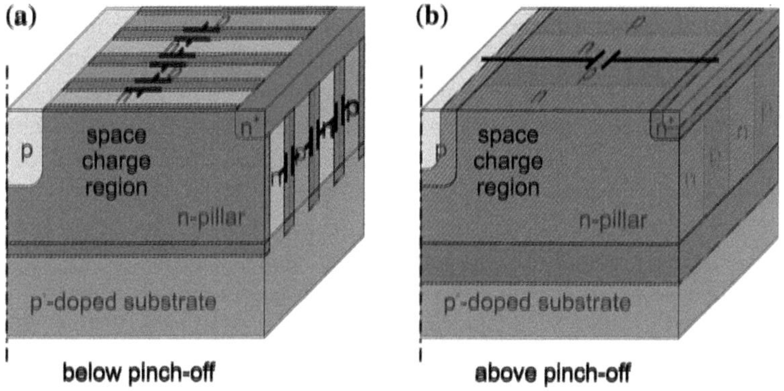

Fig. 6.20 Extension of space charge regions and output capacitances in a cells of a charge compensation pattern **a** Below and **b** Above pinch-off voltage

fully depleted resulting in a higher effective pn-junction area in comparison to RESURF-based LDMOS transistors as depicted in Fig. 6.20a.

In order to judge the impact of this pinch-off on switching performance, the output capacitance of a superjunction cell is investigated and compared to a RESURF-based drift region design. For a mainly lateral extension of the space-charge region with width

$$W_{SCR} = \sqrt{\frac{8\varepsilon_0\varepsilon_r(V_{DS} - V_{bi})}{qN_D}}, \tag{6.16}$$

the area-specific device capacitance of the charge compensation patterns is derived to

$$\frac{C_{DS,low}}{A} = \frac{\varepsilon_0\varepsilon_r}{W_{SCR}} \cdot \frac{d_{cell}}{2W_n}, \tag{6.17}$$

Here, the built-in potential for the pn-junction of the pillars obeys

$$V_{bi} = \frac{KT}{q}\ln\left(\frac{N_A N_D}{n_i^2}\right). \tag{6.18}$$

Neglecting the vertical electric field at low blocking voltages, the external voltage across the charge compensation pattern shown in Fig. 6.20 at which pinch-off in the pillars occurs is given by

$$V_{pinch} = \frac{qN_D W_N^2}{8\varepsilon_0\varepsilon_r} - V_{bi}. \tag{6.19}$$

When the blocking voltage is increased—fully depleting the pillar from free carriers—the output capacitance rapidly drops to a significantly lower value [30]. Beyond this voltage, additional charges are now influenced between p-body and n^+-drain regions across the full length of the charge compensation patterns (see right side of Fig. 6.20). This leads to a significant and steep reduction of output capacitance for high blocking voltages which does not depend on the pillar width and doping concentrations. For very high voltages, the area-specific drain-source capacitance approaches

$$\frac{C_{DS,high}}{A} = \frac{\varepsilon_0 \varepsilon_r}{2L_{cell}^2} d_{cell}. \tag{6.20}$$

Due to the interaction between lateral and vertical electric fields, this reduction does not occur instantaneously when the pinch-off voltage is exceeded. The voltage dependence of the area-specific capacitance can be estimated by

$$\left(\frac{C_{DS}}{A}(V_{DS})\right)^{-1} = \left(\frac{C_{DS,low}}{A}\right)^{-1} + \left(\frac{C_{DS,trans}}{A}(V_{DS})\right)^{-1}. \tag{6.21}$$

The last term of (6.21) was arbitrarily introduced to model the reduction of output capacitance in fabricated superjunction transistors similar to the model proposed in [31]. It is defined as

$$\frac{C_{DS,trans}}{A}(V_{DS}) = \frac{C_{DS,low} \cdot C_{DS,high}}{A\left(C_{DS,low} + C_{DS,high}\right) + \left(C_{DS,low} - C_{DS,high}\right) \cdot \exp\left(\frac{L_{cell}}{2W_n}\left(\frac{V_{DS} - V_{pinch}}{V_{pinch}}\right)\right)} \tag{6.22}$$

Note that the term has been modified to achieve a more precise fit of the area-specific capacitance for high voltages. In Fig. 6.21, the reduction of output capacitance using superjunction patterns is depicted using an analytical fitting model which was modified to account for the transition near the pinch-off [31].

From the steep reduction of output capacitance depicted in Fig. 6.21 and keeping in mind that the energy stored in a capacitor increases with the square of the capacitor voltage, a significant reduction of switching losses can be achieved in superjunction LDMOS transistors. This benefit increases for longer drift regions (increasing breakdown voltages) and smaller pillar widths and thus higher doping concentrations.

For high frequency operation, a minimization of gate-drain capacitances is desired to minimize the gate charge. As shown in Chap. 4, a source field plate is applied to conventional RF LDMOS transistors. However, facilitating this field plate in a superjunction device introduces a charge imbalance at the source-side of the charge compensation patterns due to the influence of additional charges in the p- and n-pillars, and breakdown voltage is deteriorated [25]. Consequently, a

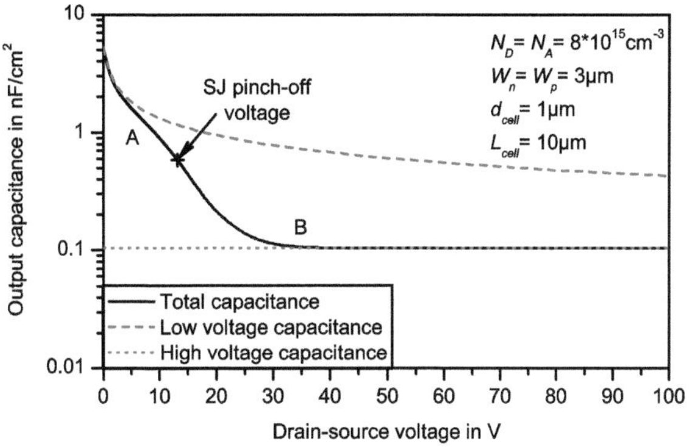

Fig. 6.21 Approximation of area-specific output capacitance of charge compensation patterns and contributions to capacitance below and above pinch-off voltage

superjunction RF LDMOS without this source field plate will introduce higher gate-drain capacitance and the transition frequency is reduced.

The introduction of a buffer-layer to compensate substrate-assisted depletion effects does not have a noticeable impact on gate charge [7]. It must be noted that the influence of superjunction patterns on the parasitic device capacitances in lateral power transistors—in general and for RF applications in particular—has received little attention so far.

6.5 Feasibility of Integration into Smart-Power ICs and RF Circuits

The reduction of drift region resistance together with the lower output capacitance make superjunction LDMOS transistors with breakdown voltages beyond 200 V an option for future high-voltage smart-power ICs. However, fabrication of suitable superjunction patterns with deep and narrowly spaced pillars imposes a severe burden towards straight forward integration. It is the realization of high aspect ratios that boosts the superjunction concept beyond RESURF-based LDMOS transistors, but CMOS-based fabrication processes do not bring alone suitable fabrication technologies. For transistors with lower breakdown voltages, the incorporation of charge compensation patterns does not help in surpassing RESURF-based LDMOS transistors for device resistance alone. In order to judge the feasibility for a specific application, additional benefits like a lower overall output capacitance need to be considered in more detail.

The incorporation of charge compensation patterns in RF LDMOS transistors using SOI technology for reduction of power losses does not appear to be feasible at the current state of technology for several reasons. Firstly, the reduction of drift region resistance compared to state-of-the-art RESURF-based RF LDMOS transistors is negligible for technologically reasonable pillar heights as concluded from Fig. 6.19 because the channel resistance of the RF LDMOS transistor is not reduced. Next, the fabrication of narrowly spaced pillar—which can be achieved in RF LDMOS technology nodes—is hampered by the interdiffusion of dopants during annealing which significantly increases drift region resistance due to the reduction of net doping in the n-pillars, and a benefit compared to RESURF-based devices could not be achieved in fabricated devices [32]. Finally, the transition frequency of superjunction RF LDMOS transistors will be lower as long as a source field plate cannot be employed. Possibly, a combination of charge compensation patterns and a RESURF region at the gate side of the drift region which allows for a source field plate could alleviate the constraints on RF performance if RF LDMOS transistors with higher blocking voltages are desired. It is for these reasons that RF LDMOS transistors with superjunction patterns have not yet been investigated.

References

1. G. Deboy, M. März, J.-P. Stengl, H. Strack, J. Tihanyi, H. Weber, New generation of high voltage MOSFETs breaks the limit line of silicon. International electron devices meeting, San Francisco, USA, 1998
2. R. Ng, F. Udrea, K. Sheng, K. Ueno, G.A.J. Amaratunga, M. Nishiura, Lateral unbalanced super junction (USJ)/3D-RESURF for high breakdown voltage on SOI. International symposium on power semiconductor devices & ICs, Osaka, Japan, 2001
3. W. Wu, B. Zhang, Z. Li, High voltage SJ-LDMOS with charge-balanced pillar and N- buffer layer. International conference on solid-state and integrated circuit technology, Xi'an, China, 2012
4. W. Wang, B. Zhang, Z. Li, Z. Li, Application of field plate in SLOP-LDMOS. International conference on communications, circuits and systems, Chengdu, China, 2010
5. T. Fujihira, Theory of semiconductor superjunction devices. Jpn. J. Appl. Phys. **36**, 6254–6262 (1997)
6. I.-Y. Park, C.A.T. Salama, CMOS compatible super junction LDMOST with N-buffer layer. International symposium on power semiconductor devices & ICs, Santa Barbara, CA, 2005
7. Y.C. Liang, G.S. Samudra, *Power Electronics—Device and Process Technologies* (World Scientific Publishing, Singapore, 2009)
8. S.G. Nassif-Khalil, L.Z. Hou, A.T. Salama, SJ/RESURF LDMOST. IEEE Trans. Electr. Dev. **51**(7), 1185–1191 (2004)
9. D. Askeland, P. Fulay, *The Science & Engineering of Materials*, 5th edn. (Thomson, Toronto, Canada, 2006)
10. M.H. Kim, J.J. Kim, Y.S. Choi, C.K. Jeon, S.L. Kim, H.S. Kang, C.S. Song, A low on resistance 700 V charge balanced LDMOS with intersected well structure. International symposium on power semiconductor devices and ICs, Cambridge, UK, 2003
11. M.J. Lin, T.H. Lee, F.L. Chang, C.W. Liaw, H.C. Cheng, Lateral superjunction reduced surface field structure for the optimization of breakdown and conduction characteristics in a high-voltage lateral double diffused metal oxide field effect transistor. Jpn. J. Appl. Phys. **42** (10), 7227–7231 (2003)
12. I.-Y. Park, A.T. Salama, Super junction LDMOS transistors. IEEE Circuits Dev. Mag. **22**, 10–15 (2006)

13. P.M. Shenoy, A. Bhalla, G.M. Dolny, Analysis of the effect of charge imbalance on the static and dynamic characteristics of the super junction MOSFET. International symposium on power semiconductor devices and ICs, Toronto, Canada, 1999
14. E. Napoli, H. Wang, F. Udrea, The effect of charge imbalance on superjunction power devices: an exact analytical solution. IEEE Electr. Dev. Lett. **29**(3), 249–251 (2008)
15. H. Wang, E. Napoli, F. Udrea, Breakdown voltage for superjunction power devices with charge imbalance: an analytical model valid for both punch through and non punch through devices. IEEE Trans. Electr. Dev. **56**(12), 3175–3183 (2009)
16. I.-Y. Park, C.A.T. Salama, Experimental implementation and characterization of a CMOS compatible buffered SJ-LDMOST. International symposium on power semiconductor and ICs, Naples, Italy, 2006
17. W. Chen, B. Zhang, Z. Li, SJ-LDMOS with high breakdown voltage and ultra-low on-resistance. Electr. Lett. **42**(22), 1314–1316 (2006)
18. M. Gross, M. Stoisiek, T. Uhlig, C. Ellmers, F. Fuernhammer, Lateral HV-MOS transistors (50 V) for integration in a 0.18 μm CMOS-process. European solid state device research conference, Munich, Germany, 2007
19. F. Udrea, A. Popescu, W.I. Milne, A new class of lateral power devices for HVIC's based on the 3D RESURF concept. Bipolar/BiCMOS circuits and technology meeting, Minneapolis, USA, 1998
20. S. Merchant, E. Arnold, H. Baumgart, S. Mukherjee, H. Pein, R. Pinker, Realization of high breakdown voltage (>700 V) in thin SOI devices. International symposium on power semiconductor devices & ICs, Baltimore, USA, 1991
21. F. Udrea, SOI-based devices and technologies for high voltage ICs. Bipolar/BiCMOS circuits and technology meeting, Boston, USA, 2007
22. S. Honarkhah, S. Nassif-Khalil, A.T. Salama, Back-etched super-juntion LDMOST on SOI, Eurpean solid-state device research conference, Leuven, Belgium, 2004
23. S.G. Nassif-Khalil, C.A.T. Salama, Super junction LDMOST on a Silicon-on-Sapphire substrate. IEEE Trans. Electr. Dev. **50**(5), 1385–1391 (2003)
24. I. Cortes, P. Fernandez-Martinez, D. Flores, S. Hidalgo, J. Rebollo, Analysis of low-voltage super-junction LDMOS structures on thin-SOI substrates. Semicond. Sci. Technol. **23**, 015009 (2008)
25. M. Bobde, L. Guan, A. Bhalla, M. Ho, Analyzing super-junction C-V to estimate charge imbalance. International symposium on power semiconductor devices & ICs, Hiroshima, Japan, 2010
26. S. Srikanth, S. Karmalkar, On the charge sheet superjunction (CSSJ) MOSFET. IEEE Trans. Electr. Dev. **55**(12), 3562–3568 (2008)
27. I. Cortes, J. Roig, D. Flores, S. Hidalgo, J. Rebollo, On the feasibility of superjunction thick-SOI power LDMOS transistors for RF base station. Semicond. Sci. Technol. **22**, 1–9 (2007)
28. A. Yoo, J.C.W. Ng, J.K.O. Sin, W.T. Ng, High performance CMOS-compatible super-junction FINFETs for sub-100 V applications. International electron devices meeting, San Francisco, USA, 2010
29. M. Rüb, M. Bär, G. Deml, H. Kapels, M. Schmitt, S. Sedlmair, C. Tolksdorf, A. Willmeroth, A 600 V 8.7Ohm mm^2 lateral superjunction transistor. International symposium on power semiconductor devices & ICs, Naples, Italy, 2006
30. S.G. Nassif-Khalil, C.A.T. Salama, 170 V super junction—LDMOST in a 0.5 μm commercial CMOS/SOS technology. International symposium on power semiconductor devices & ICs, Cambridge, UK, 2003
31. B. Zhang, J. Wu, L. Zhaoji, SLOP-LDMOS—a novel super-junction concept LDMOS and its experimental demonstration. International conferece on communications, circuits and systems, Hong Kong, China, 2005
32. M.A. Amberetu, C.A.T. Salama, 150-V Class superjunction power LDMOS transistor switch on SOI. International symposium on power semiconductors & ICs, Santa Fe, USA, 2002

Chapter 7
Lateral Power Transistors with Trench Patterns

Another emerging technology which constitutes to progress beyond modern RESURF-based LDMOS transistors utilizes trench patterns. In this chapter the trench-based technology in LDMOS transistors is reviewed. First, the applicability of trench gates is motivated by a discussion on channel resistance. The impact of FinFET technology on lateral power transistors is investigated. To evaluate the feasibility of trench gates, device designs employing trench gates and their electrical properties under static and dynamic device operation are presented and discussed. Typical Figures-of-Merit obtained from these trench devices are compared to state-of-the-art LDMOS transistors. Considering the process technology, limitations for integration into smart-power ICs and RF amplifiers are also investigated.

7.1 Contribution of Channel Resistance to Total Device Resistance in Lateral Power Transistors

In Chap. 5 on RESURF-based LDMOS transistors and Chap. 6 on superjunction LDMOS transistors, the diminishing contribution of the drift region resistance to total device resistance in LDMOS transistors with breakdown voltages below 100 V became apparent. For these devices, the contribution of channel resistance to the total device resistance described by (4.7) needs to be considered in more detail. In particular, the reduction of channel resistance is a motivation to employ trench gate devices following the concepts developed for vertical power transistors.

In order to judge the benefits of trench gate integration, a minimum channel resistance depending on breakdown voltage and technology node is derived next. Also, channel resistance is a function of the area of the cell geometry which constitutes to the density of channel regions in the device. Assuming a standard MOSFET in silicon technology, the maximum drain current in the linear region can be approximated by

T. Erlbacher, *Lateral Power Transistors in Integrated Circuits*,
Power Systems, DOI 10.1007/978-3-319-00500-3_7

$$I_D = \frac{W_{Ch}}{L_{Ch}} \mu_{n,ch} C_{ox} \left(V_{GS,max} - V_{Th} \right) U_{DS}. \tag{7.1}$$

This relationship for the channel current I_{Ch} also holds true in LDMOS transistors when the drain-source voltage is replaced by the voltage drop V_{Ch} across the channel region. Similar to (4.4), the minimum channel resistance for an LDMOS transistor at maximum gate voltage $V_{GS,max}$ is then given by

$$R_{Ch,min} = \frac{V_{Ch}}{I_{Ch}} = \frac{L_{Ch}}{W_{Ch} \mu_{n,ch} C_{ox} (V_{GS,max} - V_{Th})}. \tag{7.2}$$

Considering a standard LDMOS design with half-cell length L_{cell} (which contains exactly one channel region) and a cell width W_{cell} which equals to the channel width, an ideal area-specific channel resistance can be deduced.

$$R_{Ch,min} \cdot A = \frac{L_{Ch} L_{cell}}{\mu_{n,ch} C_{ox} (V_{GS,max} - V_{Th})}. \tag{7.3}$$

The length of the half-cell in an LDMOS transistor includes body, source and channel, drift and drain regions. Investigating state-of-the-art LDMOS transistor designs, this half-cell length can be approximated as the sum of drift region length—which depends on breakdown voltage—and 11 times the smallest feature size F, i.e.

$$L_{cell} \geqq L_{drift} + 11F \geqq \frac{BV}{E_{crit}} + 11F. \tag{7.4}$$

For (7.4) a RESURF design with a constant lateral electrical field was assumed according to (5.7). The channel length depends on the technology node used for fabrication of the transistor and is typically close to the smallest feature size F for a self-aligned gate process in silicon technology [1].

For the gate oxide capacitance, a correlation to maximum gate voltage can be found. In order to prevent oxide degradation over the lifetime of the LDMOS transistor, the electric field across the gate oxide $E_{ox,max}$ has to be limited to approximately 3 MV/cm [2]. This electric field in the oxide causes an electric field in the silicon of 1 MV/cm at the silicon/silicon dioxide interface which is well above the critical electric field for silicon. However, impact ionisation will not occur because the field is quickly reduced due to the inversion layer charge within the first 3–5 nm below the interface, and subsequently the free mean path for carriers in the inversion layer is too short to acquire enough kinetic energy required for impact ionisation. The gate oxide thickness d_{ox} is then directly determined by the maximum rated gate voltage and the electric oxide field. This gate voltage is also applied to LDMOS transistors in integrated circuits for energy conversion to minimize channel resistance. Under these considerations, the area-specific channel resistance can be rewritten as

$$R_{Ch,min} \cdot A = \frac{F\left(\frac{BV}{E_{crit}} + 11F\right)d_{ox}}{\mu_{n,ch}\varepsilon_0\varepsilon_{ox}\left(V_{GS,max} - V_{Th}\right)} = \frac{F\left(\frac{BV}{E_{crit}} + 11F\right)V_{GS,max}}{\mu_{n,ch}\varepsilon_0\varepsilon_{ox}\left(V_{GS,max} - V_{Th}\right)E_{ox,max}}, \quad (7.5)$$

with the specific dielectric constant of the gate dielectric ε_{ox}. The channel resistance exhibits a linear dependence with breakdown voltage. Neglecting the threshold voltage, which is small compared to the gate voltage, yields a channel resistance independent of the gate voltage according to

$$R_{Ch,min} \cdot A = \frac{F\left(\frac{BV}{E_{crit}} + 11F\right)}{\mu_{n,ch}\varepsilon_0\varepsilon_{ox}E_{ox,max}}. \quad (7.6)$$

This consideration includes that a higher gate voltage generates more inversion layer charges but a higher gate oxide thickness has to be used to withstand this gate voltage. Moreover, using the dependence of the critical electric field on breakdown voltage from (4.24), channel resistance is solely determined by material parameters and the smallest feature size:

$$R_{Ch,min} \cdot A = 3861 \cdot F\left(\frac{BV^{\frac{7}{6}}}{7.95 \cdot 10^5} + 11 \cdot F\right), \quad (7.7)$$

Note that the inversion layer mobility like the bulk mobility depends on the doping concentration. Additionally, under high vertical electric fields, the inversion layer mobility is significantly reduced. Models for calculation of the inversion layer mobility under high vertical electric fields have been previously reported [3, 4]. Regarding the investigation on ideal channel resistance in LDMOS transistors, an inversion layer mobility of approximately 250 cm^2/Vs is predicted according to these models when a local electric field of 1 MV/cm is assumed.

Based on the above stipulations, the ideal channel resistance limit for a planar LDMOS device for different technology nodes has been summarized and compared to the drift region resistances for the 1D punch-through case and a RESURF design in Fig. 7.1.

It is evident that a reduction of total device resistance of integrated power semiconductor devices in the voltage below 100 V can be achieved by a further reduction of channel resistance. The benefit of using fabrication technologies with smaller feature sizes is clearly visible for breakdown voltages below 50 V. For higher voltages, the drift region resistance becomes dominant as discussed in Chaps. 4–6.

The breakdown voltage at which the transition from channel region to drift region limited current conduction takes place in RESURF-based LDMOS transistors is decreasing for devices fabricated with smaller feature sizes. However, the reduction of channel length for LDMOS transistors in integrated circuits by progressive scaling of smallest feature size (More-Moore approach) is not readily

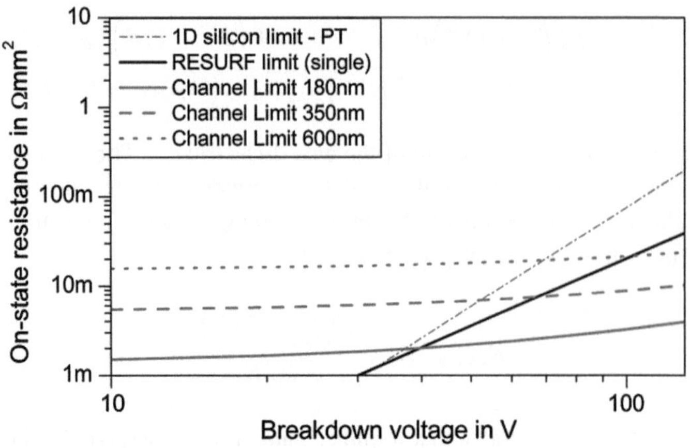

Fig. 7.1 Ideal channel resistance in different technology nodes and drift region limits for planar LDMOS transistors

available because processing technologies for smart-power ICs are lagging behind and depend on future advances in CMOS processing. Therefore, other means of reducing channel resistance following the More-than-Moore approach have been exploited.

7.2 LDMOS Device Designs for Smart-Power ICs Utilizing Trench Patterns

Several concepts based on trench gates have been investigated in order to achieve a significant reduction of channel resistance in lateral power transistors for integrated circuits:

- Utilizing semiconductor fins increases the effective channel width by providing lateral channel regions along sidewalls of the etched semiconductor device.
- Implementation of trench gates while maintaining a lateral drift region provides channel regions with a vertical gate current into the semiconductor device. This allows for a higher integration density similar to the trench design used in vertical power transistors (e.g. UMOS) [5].
- Incorporation of trench gates with a vertical drift region achieves highest integration density. Additional processing steps are required to realize a lateral device design by collecting the current of these quasi-vertical transistors and routing it back to the semiconductor surface.

These topologies are reviewed in the following paragraphs.

7.2.1 Increased Channel Width Using FinFET Topology

An increase in effective channel width has been successfully demonstrated for logic level MOSFETs by extending the channel regions into the third dimension, i.e. perpendicular to the semiconductor surface [6]. Fin-shaped MOSFET transistors (FinFETs) that are based on this concept have received significant attention in highly integrated CMOS processes [7]. By transferring this idea to LDMOS transistors, FinFET-type device structures as illustrated in Fig. 7.2 have been developed.

This FinFET topology provides additional inversion channels at the sidewalls of each silicon fin, and effective channel resistance is lowered. For LDMOS transistors using these lateral FinFET gate structures a breakdown voltage of 27.4 V at an area-specific device resistance of 46 mΩ mm^2 was predicted by TCAD simulations—exceeding the 1D channel limit for a 350 nm technology.

Implementation of FinFET patterns into LDMOS transistors using 400 nm wide silicon trenches with 1 μm depth was experimentally verified in a geometry given in Fig. 7.3.

While trench etching requires some additional effort, the drift region layout of a typical LDMOS transistor can be maintained. Electrical measurements on these devices revealed a breakdown voltage of 25 V at a device resistance of 130 m Ω mm^2, which is close to the 1D channel limit for an 180 nm technology and demonstrates the capability of this approach.

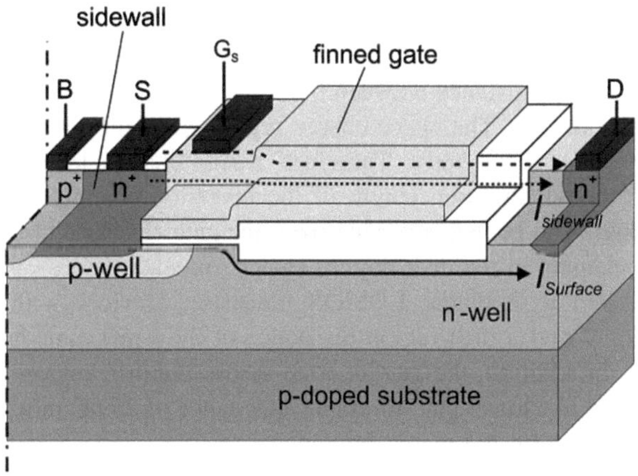

Fig. 7.2 FinFET-type LDMOS transistor concept using lateral finned gates, adopted from [8]

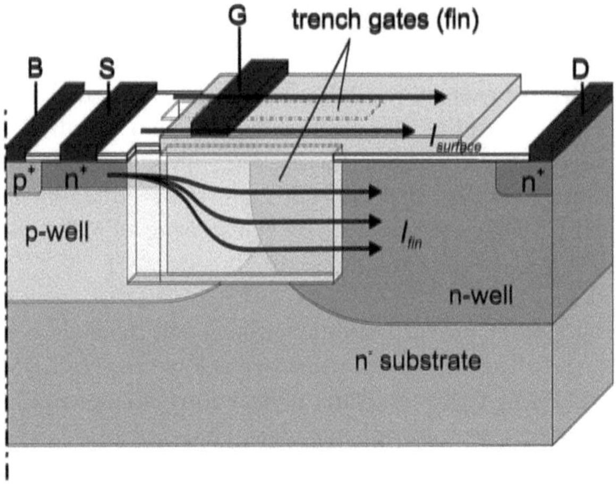

Fig. 7.3 Topology of fabricated FinFET-type LDMOS transistor using lateral finned gates, adopted from [9]

7.2.2 Shallow Trench Isolation for Trench Gate Transistors

Availability of three-dimensional gate geometries enabled a broad range of novel device designs. Incorporation of this technology into smart-power IC processes was demonstrated for LDMOS transistors using standard shallow trench isolation (STI). A device concept which does not require additional trench etching effort is presented in Fig. 7.4.

This transistor design uses the STI trench to implement a vertical channel portion. By doing so, the distance between the n^+-source and the drift region is larger than in a planar device. The space-charge region from the drift region does not extend as far across the p-body. Therefore, a shorter gate length can be implemented. Even though the total length of the inversion channel is not reduced, a smaller cell length can be realized. Moreover, the gate electrode also creates a local accumulation channel in the drift region. Hence, on-state device resistance can be reduced compared to a planar LDMOS transistor. Devices with a breakdown voltage of 35.2 V and a drift region resistance of 32.4 mΩ mm^2 have been demonstrated [10]. In addition, the gate overlap across the drift region is significantly reduced, resulting in a lower gate-to-drain capacitance of 25 nC/mm^2 compared to a value of 40 nC/mm^2 for the planar reference.

Using a similar design which extends the planar gate electrode into an STI-shaped trench in the drift region predicted a drift region resistance of 70 V at an on-state resistance of 70 mΩ mm^2 [11].

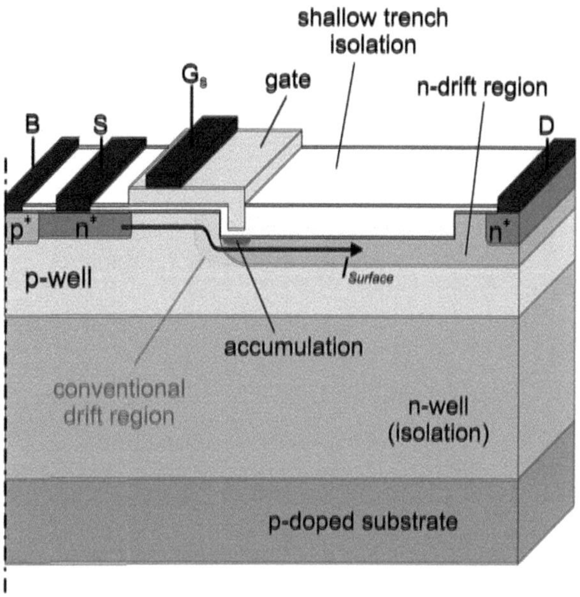

Fig. 7.4 LDMOS transistor concept with a gate trench formed in the STI, adopted from [10]

7.2.3 Vertical Channels Using Trench Gates

In vertical power semiconductor devices, vertical trench gates were introduced as a method for increasing the number of channels per area. Similar to the FinFET concept, vertical trench gates have also been investigated with respect to LDMOS transistors. A straight-forward approach offering high integration density is presented in Fig. 7.5.

The footprint for the gate electrode is significantly reduced. At the same time, a standard RESURF-based drift region design of conventional LDMOS technology can be retained. The presented design also provides a full depletion of the drift region between substrate and p-well at low voltages, shielding the gate electrode from high drain voltages and suppressing hot carrier injection [12]. This vertical depletion effect also reduces transfer and output capacitances compared to planar LDMOS transistors. Fabricated lateral trench gate power transistors with a breakdown voltage of 36 V exhibit a low on-state resistance of 90 mΩ mm^2.

A similar trench gate transistor design shown in Fig. 7.6 emphasizes the idea of reducing channel resistance by trench gate integration in order to construct devices with on-state resistance close to the drift region limit even for breakdown voltages below 100 V.

Here, a double-acting RESURF was combined with both trench gate and trench drain topologies. Thereby, the current injected through the channel into the drift region below the silicon surface is enhanced by a low-ohmic drain contact, i.e. the

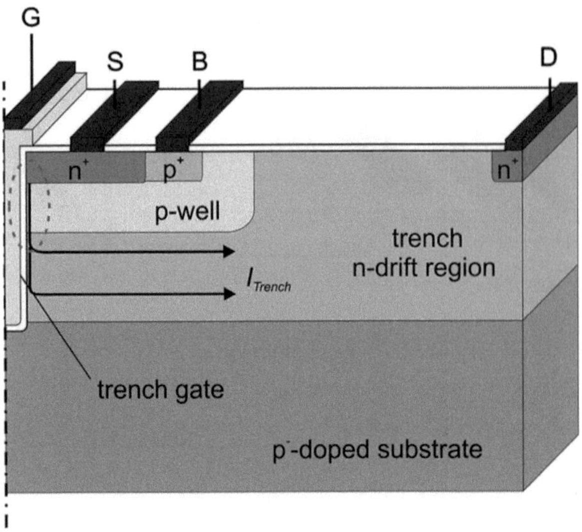

Fig. 7.5 Trench-gate LDMOS transistor concept with planar drain, adopted from [12]

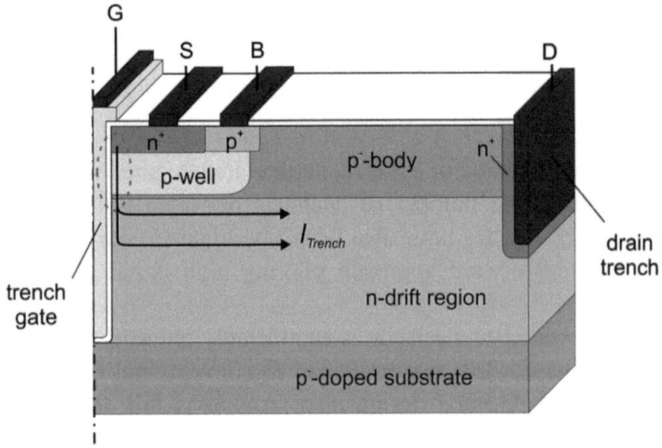

Fig. 7.6 Trench-gate LDMOS transistor concept with drain trench, adopted from [13]

current through the drift region is parallel to the silicon surface and the effective drift distance for the electrons is minimized. TCAD simulations indicate that on-state device resistances of 41.4 and 130 mΩ mm^2 could be achieved in such an LDMOS transistor with breakdown voltages of 56.4 and 104 V, respectively. Of course, incorporation of the drain trench electrode requires additional fabrication effort.

7.2.4 Quasi-vertical Transistors Using Trench Gates

Taking the integration of trench gates one step further, quasi-lateral power transistors with a vertical drift region allow for even lower drift region resistances [14]. Despite the additional fabrication effort, the device concept presented in Fig. 7.7 is applicable to smart-power IC processing for devices with breakdown voltages up to 150 V.

This geometry requires a highly n-doped buried layer and sinkers at the boundaries of the transistor cells to minimize series resistance of the drain contact. The trench gate is located close to the surface of the transistor. Beneath, the drift region is implemented by a superjunction-like geometry of n-pillars and the deep trenches filled with oxide and polysilicon. This capacitor acts like a highly doped p-layer in forward blocking operation, fully depleting the n-pillars at a low voltage. Due to the tight spacing, a high n-pillar doping concentration can be used (see Chap. 6). The combination of high integration density of channels and highly doped drift regions allows for an impressive reduction of drift region resistance. An experimentally verified device resistance of 30 mΩ mm^2 for such a quasi-vertical LDMOS transistor with a breakdown voltage of 94 V fabricated in a 0.35 μm CMOS technology is close to the theoretical RESURF-based drift region limit [15]. However, the large gate-drain overlaps along the pillars and the high integration density results in high gate charge during switching operation of 4.8 nC/mm^2.

7.2.5 Trench Gate Designs Beyond CMOS Technology

When the use of standard CMOS technology is not a constraint, more complex trench gate patterns can be employed to further improve device performance. For

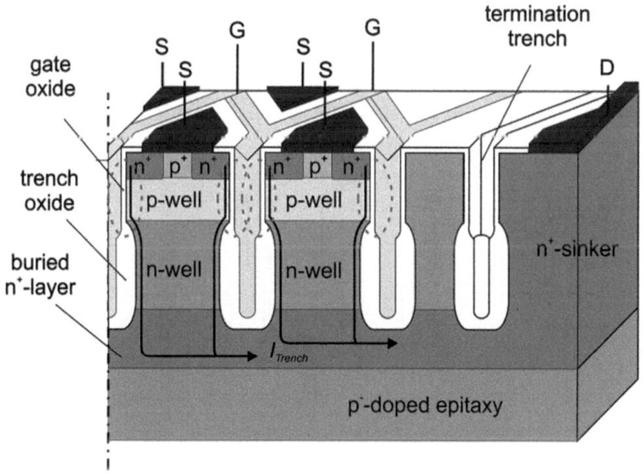

Fig. 7.7 Quasi-vertical trench-gate LDMOS transistor concept, adopted from [14]

example, a trench gate LDMOS transistor with an additional buried gate has been proposed as shown in Fig. 7.8.

This device is tailored to minimize device resistance by implementation of two conductive paths at the trench gate and at the buried gate. Measured devices exhibit a breakdown voltage of 49.5 V and an on-state resistance of 42 mΩ mm^2. However, fabrication requires epitaxial overgrowth and subsequent thinning which are both not readily available in current commercial semiconductor processing technologies.

Another highly integrated LDMOS device concept with buried-source which also aims at low coupling capacitances is shown in Fig. 7.9.

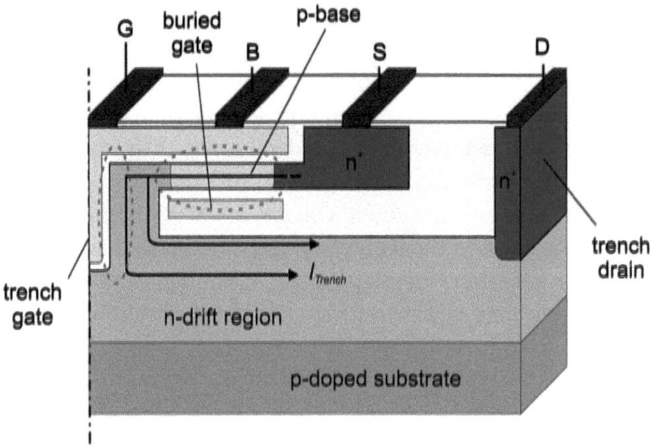

Fig. 7.8 Trench-gate LDMOS transistor concept with buried gate, adopted from [16]

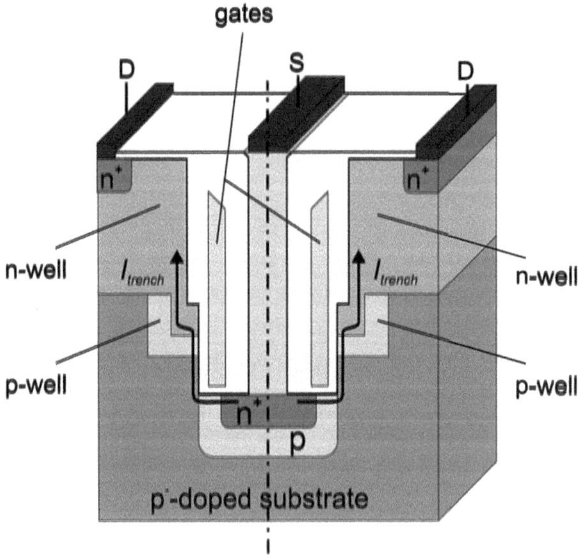

Fig. 7.9 Quasi-vertical trench-gate LDMOS transistor concept, adopted from [17]

Here, the drain contact is placed at the bottom of the trench which also houses the vertical gate electrodes. A significant reduction of device resistance can be realized due to the high integration density. Investigated samples provided an on-state resistance of 69.2 mΩ mm^2 at a breakdown voltage of 96 V. However, a gate charge during switching of 10 nC/mm^2 is incurred due to the high integration density.

7.2.6 *Comparison of Trench Pattern Designs for Power Electronic Applications*

A summary of relevant device parameters for different LDMOS transistors based on trench gates presented in literature is listed in Tables 7.1 and 7.2.

The data presented here will be used to compare the different trench gate designs with state-of-the-art LDMOS transistors in Sect. 7.4.

7.3 RF LDMOS Device Designs Utilizing Trench Patterns

Integration of trench gates has also received attention for integrated RF LDMOS transistors. For LDMOS transistors with breakdown voltages below 60 V that are used in state-of-the-art RF applications, the drift region resistance is negligible in present CMOS technologies. This allows for a certain degree of freedom in the transistor design. A trench LDMOS design which can be implemented into standard CMOS technology with manageable effort is shown in Fig. 7.10 [23].

Table 7.1 Breakdown voltage, on-state resistance and gate charge measured from fabricated LDMOS transistors using trench gates reported in literature

BV (V)	$R_{DS,on}$ (Ω mm^2)	Q_{gd} (nC/mm^2)	Device type	Substrate	Source
25	0.013	–	FinFET LDMOS	Bulk Si	[9]
35.2	0.0324	25	TG-LDMOS	Bulk Si	[10]
37	0.09	–	TG-LDMOS	Bulk Si	[12]
94	0.03	4.8	QV TG-LDMOS	Bulk Si	[15]
72	0.06	–	TG-LDMOS	Bulk Si	[17]
96	0.0692	10	TG-LDMOS	Bulk Si	[17]
11.6	0.0106	–	TG-RF LDMOS	Bulk Si	[18]
49.5	0.042	–	TG-LDMOS	Bulk Si	[16]
80	0.08	–	TG-LDMOS	Bulk Si	[19]
26	0.01	1.5	TG-LDMOS	Bulk Si	[20]

Table 7.2 Breakdown voltage, on-state resistance and gate charge from simulations on LDMOS transistors using trench gate topology reported in literature

BV (V)	$R_{DS,on}$ (Ω mm^2)	Q_{gd} (nC/mm^2)	Device type	Substrate	Source
27.4	0.0046	–	FinFET LDMOS	Bulk Si	[8]
25	0.0078	–	FinFET LDMOS	Bulk Si	[9]
70	0.07	–	STI LDMOS	Bulk Si	[11]
56.4	0.0414	–	TG LDMOS	Bulk Si	[13]
104	0.13	–	TG LDMOS	Bulk Si	[13]
91.5	0.117	1.133	CTGI LDMOS	Bulk Si	[21]
90.6	0.094	1.81	ITGI LDMOS	Bulk Si	[21]
84	1.66	–	TG-SJ LDMOS	SOI	[22]

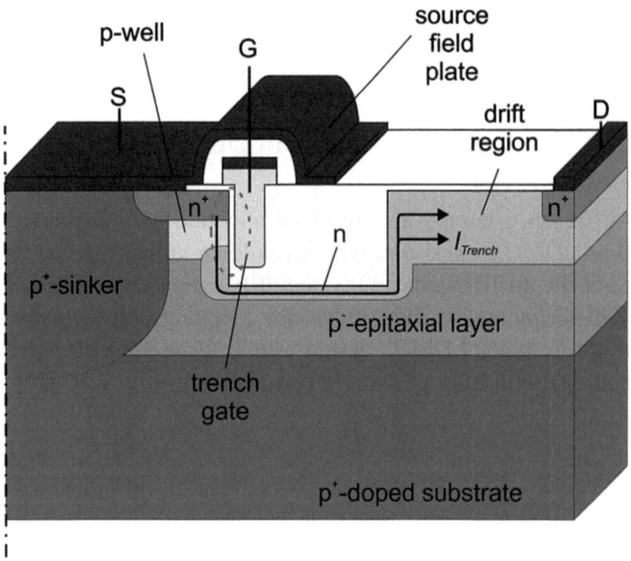

Fig. 7.10 Trench gate LDMOS transistor design for RF applications, adopted from [23]

Moving the channel region into the silicon substrate by such a device design allows reducing the electric field on the semiconductor surface in the drift region. Then, the gate field plate which is present to repel the electric field from the surface of the drift region in planar power transistors can be omitted, and the gate-to-drain capacitance is reduced. As this capacitance significantly impacts RF performance, trench gate integration provides a promising means to implement RF LDMOS devices. To further minimize gate-to-drain capacitance, the trench gate can be implemented asymmetrically. Here, a thin gate oxide at the channel region and a thicker dielectric towards the drift layer is used. Besides providing less coupling capacitance, the latter also achieves sufficient voltage stability between the gate dielectric and the drift region. Overall, this technology could enable the integration of RF LDMOS transistors with a transition frequency of 6 GHz at a breakdown voltage of 96 V [23].

7.3.1 Shallow Trench Isolation

Aside from using trench gates, RF LDMOS transistors with other trench structures have been demonstrated. Performance of RF LDMOS devices can be improved by employing the shallow trench isolation pattern of CMOS technology between gate and drain [18]. Figure 7.11 depicts such a device structure with an operation voltage of 11.6 V.

By carefully selecting the width, depth and offset distance from the gate, an on-state resistance of 3.8 Ω mm was achieved. Due to the placement of the drift region below the semiconductor surface, the electric field at the surface is limited to acceptable values even without a gate field plate. This allows for a low gate-to-drain capacitance and enables decent RF operation. A transition frequency of 30 GHz and a power added efficiency measured under class AB operation at 2.45 GHz of 55 % were reported.

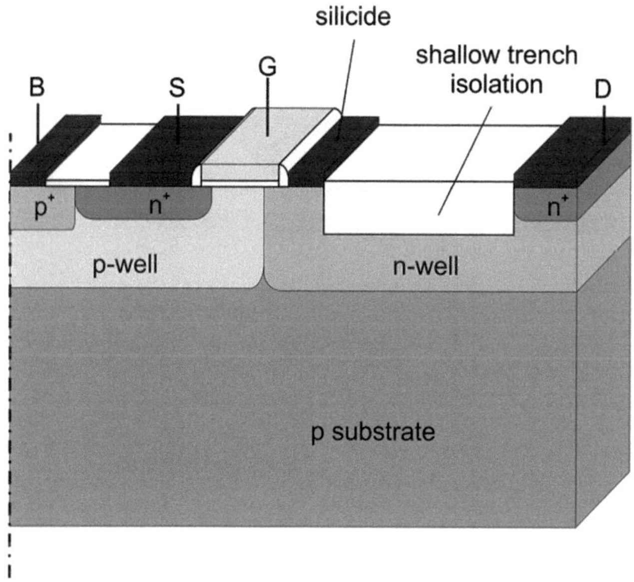

Fig. 7.11 RF LDMOS with shallow trench isolation in the drift region, concept adopted from [18]

7.3.2 Trench Sinker for Cell Length Reduction

In conventional RF LDMOS transistors, the substrate is biased to source potential by a p^+-sinker (see Chap. 4). Ordinarily, this sinker is fabricated by a deep diffusion process which also causes lateral extension of the sinker similar to its depth. Especially for RF transistors, where the drift region length is small (breakdown voltages below 60 V), this diffused sinker causes a major contribution to cell pitch. By implementing a trench-based sinker which can be filled by highly doped polysilicon as illustrated in Fig. 7.12, a low-ohmic connection between the substrate and the source electrode can be fabricated.

Fabricated transistors using this technology achieved a breakdown voltage of 70 V at an on-state resistance of 3.7 Ω mm and a transition frequency of 9 GHz. A maximum output power of 30 dBm with a power gain of 15.8 dB was measured at a frequency of 2 GHz on wafer level.

7.3.3 Comparison of Trench Pattern Designs for RF Applications

Table 7.3 contains a summary of device properties of proposed RF LDMOS transistors employing trench gate technology.

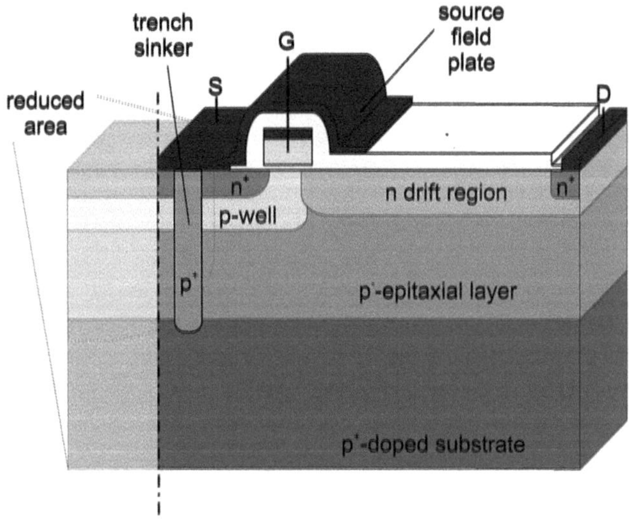

Fig. 7.12 RF LDMOS with trench sinker, concept adopted from [24]

Table 7.3 Breakdown voltage, cut-off and maximum oscillation frequencies for RF LDMOS transistors using trench gates reported in literature

BV (V)	f_T (GHz)	f_{max} (GHz)	Device type	Substrate	Source
11.6	18	30	STI RF LDMOS	Bulk Si	[18]
11.6	18	30	STI RF LDMOS	Bulk Si	[18]
98	–	6	TG RF LDMOS	Bulk Si	[23]
70	–	9	TS RF LDMOS	Bulk Si	[24]
70	–	9	TS RF LDMOS	Bulk Si	[24]
15	18	27	STI RF LDMOS	Bulk Si	[25]

7.4 Electrical Properties Under Static and Dynamic Device Operation

For smart-power ICs, a reduction of area-specific device resistance is often desired when implementing devices with trench patterns. This allows for a reduction in chip size for a given power demand in an existing IC layout or the evolution of ICs with high power density. The efficiency of integrating trench gates is summarizes in Fig. 7.13 as a comparison of breakdown voltage and on-state resistance to state-of-the-art RESURF-based LDMOS transistors.

Some of the introduced designs with trench patterns achieve a significant reduction of device resistance. This is especially true in the voltage range between 60 and 100 V where the transition from channel to drift region dominated device resistance occurs. Despite the potential of low-loss trench-based LDMOS transistors at voltages below 50 V indicated by simulations, such devices have not yet

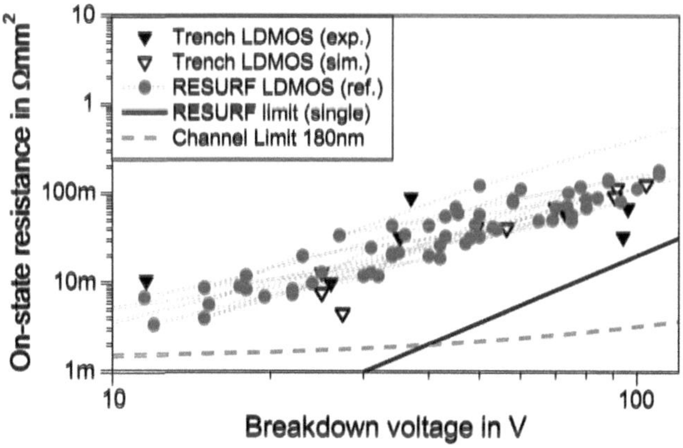

Fig. 7.13 Total device resistance versus breakdown voltage for trench-gate LDMOS transistors compared to standard LDMOS transistors, data from Tables 7.1 and 7.2

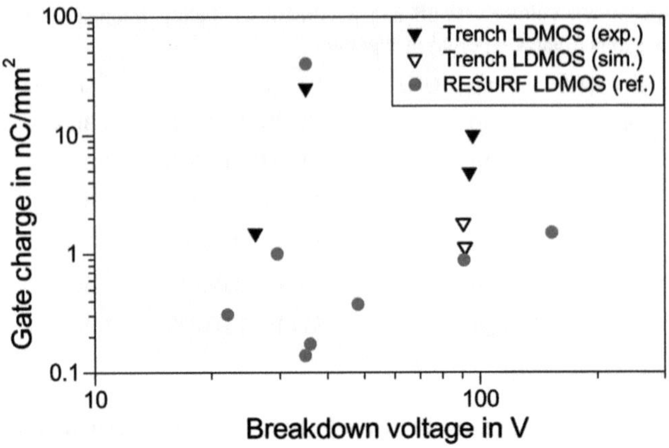

Fig. 7.14 Gate charge versus breakdown voltage for trench-gate LDMOS transistors compared to standard LDMOS transistors, data from Tables 7.1 and 7.2

been fabricated in a technology that could be implemented in CMOS technology. In fact, at such voltages, the 1D channel limit is yet to be overcome.

With respect to higher operating frequencies in smart-power ICs, a reduction of gate charge as a measure for dynamic losses has been achieved by trench gate integration. Figure 7.14 illustrates the comparison between trench-based and RESURF-based LDMOS transistors for different breakdown voltages.

A key concept in trench based LDMOS transistors is the ability to omit the gate field plate which significantly contributes to gate-to-drain capacitance and increases dynamic switching losses in smart-power ICs. Concerning RF LDMOS transistors in integrated circuits, a comparison of transition frequencies for different device designs is illustrated in Fig. 7.15.

An increase in transition frequencies was demonstrated for CMOS-compatible trench-based LDMOS transistors compared to planar devices. While the benefit appears negligible for voltages below 50 V, transistors with significantly higher transition frequency were demonstrated at voltages close to 100 V. It should be emphasized that the omission of the gate field plate by trench integration reduces coupling capacitances.

7.5 Limitations and Feasibility of Integration into Smart-Power ICs and RF Circuits

As with every integration concept that relies on a modification of the process flow, most trench gate integration designs require additional process steps and mask layers—at least for patterning the trench. On one hand, trench gate LDMOS transistors exhibit lower on-state resistance at a given breakdown voltage due to the

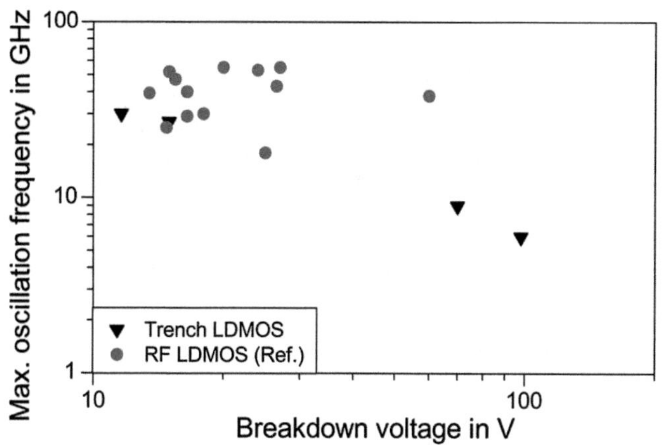

Fig. 7.15 Maximum oscillation frequency versus breakdown voltage for trench-gate RF LDMOS compared to planar RF LDMOS transistors, data from Table 7.3

higher integration density of channels. On the other hand, this high channel density also gives rise to higher gate charge values. This limits their applicability to low switching frequencies in general and care must be taken to provide the precise amount of channel density for a device designed for a given voltage rating. For superior RF performance and low dynamic losses, more sophisticated geometries need to be implemented. This also increases the fabrication effort and requires outstanding performance for applications with enough market volume to be implemented. As efficient planar RF LDMOS transistors are available for amplifiers in mobile communication (up to 5 GHz), trench gate integration may not offer significant benefits to warrant its implementation in volume production. Nevertheless, trench based RF LDMOS transistors could be an enabling technology to realize RF amplifiers with high output power due to application of operating voltages of 50 V and beyond. To this end, sufficient investigations of RF LDMOS technology based on trench gates have not yet been published.

References

1. D.J. Foster, in *Electronic Materials*, Silicon Processing: CMOS Technology (Springer, New York, 1991), pp. 173–191
2. D.J. Dumin, *Oxide Reliability: A Summary of Silicon Oxide Wearout, Breakdown, and Reliability* (World Scientific, Singapore, 2002)
3. S.C. Sun, J.D. Plummer, Electron mobility in inversion and accumulation layers on thermally oxidized silicon surfaces. IEEE Trans. Electron Devices **ED-27** (8), 1497–1508 (1980)
4. S. Takagi, A. Toriumi, M. Iwase, H. Tango, On the universality of inversion layer mobility in Si MOSFET's: Part I-effects of substrate impurity concentration. IEEE Trans. Electron Devices **41**(12), 2357–2362 (1994)

5. R.S. Saxena, M.J. Kumar, in *Trench Gate Power MOSFET: Recent Advances and Innovations*, ed. by S. Jit. Advances in Microelectronics and Photonics (Nova Science Publishers Inc., New York, 2012), pp. 1–23
6. X. Huang, W.-C. Lee, C. Kuo, D. Hisamoto, L. Chang, J. Kedzierski, E. Anderson, H. Takeuchi, Y.-K. Choi, K. Asano, V. Subramanian, T.-J. King, J. Bokor, C. Hu, Sub 50-nm FinFET: PMOS, in *International Electron Devices Meeting*, Washington, USA, 1999
7. D. Hisamoto, W.-C. Lee, J. Kedzierski, H. Takeuchi, K. Asano, C. Kuo, E. Anderson, T.-J. King, J. Bokor, C. Hu, FinFET—A self-aligned double-gate MOSFET scalable to 20 nm. IEEE Trans. Electron Devices **47**(12), 2320–2325 (2000)
8. B. Duan, Y. Yang, B. Zhang, X. Hong, Folded-accumulation LDMOST: new power MOS. IEEE Electron Device Lett. **30**(12), 1329–1331 (2009)
9. A. Nakagawa, Y. Kawaguchi, Improved 20 V lateral trench gate power MOSFETs with very low on-resistance of 7.8 mΩ mm^2, in *International Symposium on Power Semiconductor Devices and ICs*, Toulouse, France, 2000
10. H. Wang, H.P.E. Xu, W.T. Ng, K. Fukumoto, K. Abe, A. Ishikawa, Y. Furukawa, H. Imai, T. Naito, N. Sato, K. Sakai, S. Tamura, K. Takasuka, A novel orthogonal gate EDMOS transistor with improved dv/dt capability and figure of merit (FOM). IEEE Electron Device Lett. **29**(12), 1386–1388 (2008)
11. M. Zitouni, F. Morancho, P. Rossel, H. Tranduc, J. Buxo, I. Pagès, A new concept for the lateral DMOS transistor for smart power IC's, in *International Symposium on Power Semiconductor Devices and ICs (ISPSD)*, Toronto, Canada, 1999
12. D. Disney, W. Chan, R. Lam, R. Blattner, S. Ma, W. Seng, J.-W. Chen, M. Cornell, R. Williams, 60 V lateral trench MOSFET in 0.35 μm technology, in *International Symposium on Power Semiconductor Devices and ICs*, Orlando, USA, 2008
13. I.-Y. Park, Y.-I. Choi, S.-K. Chung, H.-J. Lim, S.-I. Mo, J.-S. Choi, M.-K. Han, Numerical analysis on the LDMOS with a double epi-layer and trench electrodes. Microelectron. J. **32**, 497–502 (2001)
14. P. Moens, F. Bauwens, J. Baele, K. Vershinin, E. De Backer, E. M. Sankara Narayanan, M. Tack, XtreMOS: the first integrated power transistor breaking the silicon limit, in *International Electron Device Meeting*, San Francisco, USA, 2006
15. P. Moens, F. Bauwens, B. Desoete, J. Baele, K. Vershinin, H. Ziad, E. M. Shankara Narayanan und M. Tack, Record-low on-resistance for 0.35 μm based integrated XtreMOS transistors, in *International Symposium on Power Semiconductor Devices and ICs*, Jeju, Korea, 2007
16. T. Uesugi, M. Kodama, S. Kawaji, K. Nakashima, Y. Murase, E. Hayashi, Y. Mitsushima, H. Tadano, New 3-D lateral power MOSFETs with ultra low on-resistance, in *International Symposium on Power Semiconductor Devices and ICs*, Kyoto, Japan, 1998
17. N. Fujishima, A. Sugi, S. Kajiwara, K. Matsubara, Y. Nagayasu, C.A.T. Salama, A high-density low on-resistance trench lateral power MOSFET with a trench bottom source contact. IEEE Trans. Electron Devices **49**(8), 1462–1468 (2002)
18. T. Yan, H. Liao, Y.Z. Xiong, R. Zeng, J. Shi, R. Huang, Cost-effective integrated RF power transistor in 0.18-μm CMOS technology. IEEE Electron Device Lett. **27**(10), 856–858 (2006)
19. N. Fujishima, C.A.T. Salama, A trench lateral power MOSFET using self-aligned trench bottom contact holes, in *International Electron Devices Meeting*, Washington, USA, 1997
20. S.T. Peake, R. Grover, R. Farr, C. Rogers, G. Petkos, Fully self-aligned power trench-MOSFET utilising 1 μm pitch and 0.2 μm trench width, in *International Symposium Power Semiconductor Devices and ICs*, Santa Fe, USA, 2002
21. T. Erlbacher, A. J. Bauer, L. Frey, Significant on-resistance reduction of LDMOS devices by intermitted trench gates integration. IEEE Trans. Electron Devices **59**, 3470–3476 (2012)
22. I. Cortes, P. Fernandez-Martinez, D. Flores, S. Hidalgo, J. Rebollo, Analysis of low-voltage super-junction LDMOS structures on thin-SOI substrates. Semicond. Sci. Technol. **23**, 015009 (2008)
23. P.H. Wilson, A novel trench gate LDMOS for RF applications, in *International Crimean Conference on Microwave and Telecommunication Technology*, Sevastopol, Ukraine, 2003

24. C.S. Kim, J.-W. Park H.K. Yu, Trenched sinker LDMOSFET (TS-LDMOS) structure for high power amplifier applications above 2 GHz, in *International Electron Devices Meeting*, Washington, USA, 2001
25. H. Xiao, L. Zhang, R. Huang, F. Song, D. Wu, H. Liao, W. Wong, Y. Wang, A novel RF LDMOS fabricated with standard foundry technology. IEEE Electron Device Lett. **30**(4), 386–388 (2009)

Chapter 8
Lateral Power Transistors Combining Planar and Trench Gate Topologies

Progress in the development of novel LDMOS transistors in integrated circuits has also been made by combination of planar and trench gate topologies. This effort is discussed as an example of novel device technologies in more details. Based on the trench gate topologies described in Chap. 7, the combination of existing planar LDMOS designs with trench gate topology appeals from its ease of integration into commercially available smart-power IC fabrication processes. In this light, the electrical characteristics under static and dynamic device operation are discussed. Due to the device design, avalanche ruggedness and limitations under operation as high-side switches are especially considered. The resulting FOMs are analysed with respect to implementation of these devices for high output power and high frequency operation.

8.1 Device Concepts for Combination of Planar and Trench Gates

Given that there are already a large number of industrialized smart-power IC fabrication technologies available, integration concepts allowing for the utilization of established topologies like RESURF designs and field plates are especially desirable. The trench gate integration concepts shown in Chap. 7 promise a significant reduction of device resistance. However, the device topologies cannot be readily implemented due to different wells and geometries especially required for the trench gate integration approach [1]. At least, a redesign of cell geometries is necessary in the case of the superjunction technology to utilize the advantages of a higher drift region doping [2].

8.1.1 Continuous Trench Gate Integrated LDMOS

Hence, integration of trench gates into existing planar LDMOS topology offers promising perspectives by inheriting the advantages of both trench gate and RESURF concepts [3] while providing ease of integration into existing process flows. This concept represents a true "More-than-Moore" integration approach.

T. Erlbacher, *Lateral Power Transistors in Integrated Circuits*,
Power Systems, DOI 10.1007/978-3-319-00500-3_8

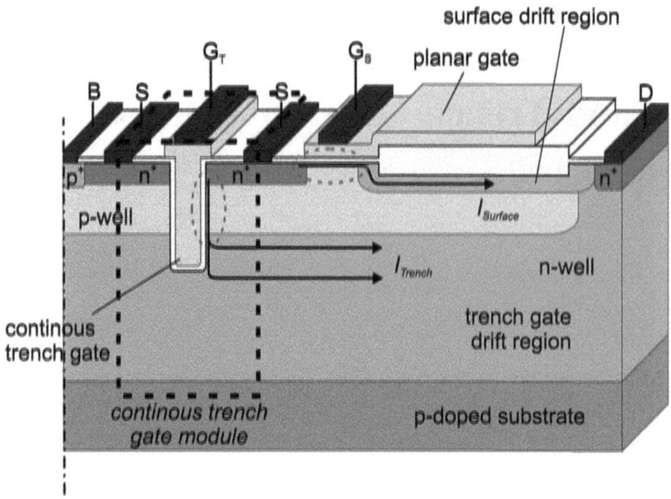

Fig. 8.1 Trench gate integrated LDMOS transistor (with extended half-cell length) exhibiting two different current paths from source to drain—additional trench gate module is indicated by *dashed lines*

The cross-section of such a trench gate integrated LDMOS device with extended half-cell length is schematically depicted in Fig. 8.1.

The *Trench Gate Integrated LDMOS* (TGI-LDMOS) consists of a standard planar LDMOS cell with RESURF design and a field plate. In addition, a trench gate module has been added at the source side of the LDMOS intersecting the n^+-source regions. Hereby, the trench gate vertically penetrates through the p-well (body of the LDMOS transistor), making the n-well region accessible for current conduction by two vertical channel regions per trench. In the planar LDMOS design, this trench drift region is merely used to block the drain voltage against the substrate. As such it is already designed to support the full drain-source blocking voltage featuring a double acting RESURF. Consequently, the double-diffused regions of the underlying LDMOS transistor can remain unchanged for sound operation of the TGI-LDMOS. This minimized the number of additional process steps required for its fabrication, ensuring straight-forward integration into foundry processes, where the well design is already determined by the CMOS and planar LDMOS processes [4]. This also provides the opportunity to manufacture planar LDMOS devices and TGI-LDMOS transistors on the same chip.

Even though it is not the preferred method of operation in Smart-Power ICs, it should be noted that the conduction of planar and trench gate currents through this type of devices can be independently controlled by the planar and trench gates. This gives rise to the application of the TGI-LDMOS as a highly integrated trinary switch [5].

The extension of the planar LDMOS by the trench gate module partially diminishes the benefit of a higher drain current at the expense of a larger cell footprint. To further increase integration density and reduce the area-specific device

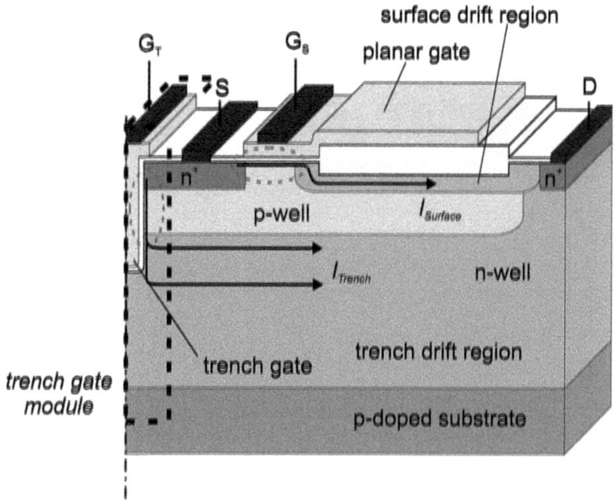

Fig. 8.2 Trench gate integrated LDMOS transistor (half-cell) employing an area neutral integration concept—additional trench gate module is indicated by *dashed lines*

resistance, an area-neutral TGI-LDMOS concept can be applied. The corresponding cross-section is presented in Fig. 8.2.

There, the p-contact region is replaced by the trench gate module which results in a significantly lower cell length. In turn, the on-state resistance can be minimized by placing more cells on the chip. However, the contacts to the body region cannot be fully omitted and need to be placed at the boundaries of the p-wells between adjacent cells in order to prevent parasitic bipolar effects [6]. Therefore, this periodic placement of the body contacts along the width of the device imposes a limit on the cell width with respect to switching frequency and parasitic bipolar turn-on. Note that this approach is also not ideally suited for low-standby power applications where body switching is employed to minimize leakage currents [7].

8.1.2 Intermitted Trench Gate Integrated LDMOS

Further reduction of drift resistance can be achieved by implementation of intermitted trench gates rather than continuous trench gates [8]. The intermitted trench gate integrated device concept is illustrated in Fig. 8.3.

The key advantage compared to a continuous trench gate is a reduction of cell length while maintaining the body contact region. This is achieved by sharing one source contact for both trench and planar gates. Then, the source current for the planar gate can pass between the intermitted trench gates. This is also depicted in the schematic top view of Fig. 8.4 for the intermitted trench gate integrated LDMOS transistor.

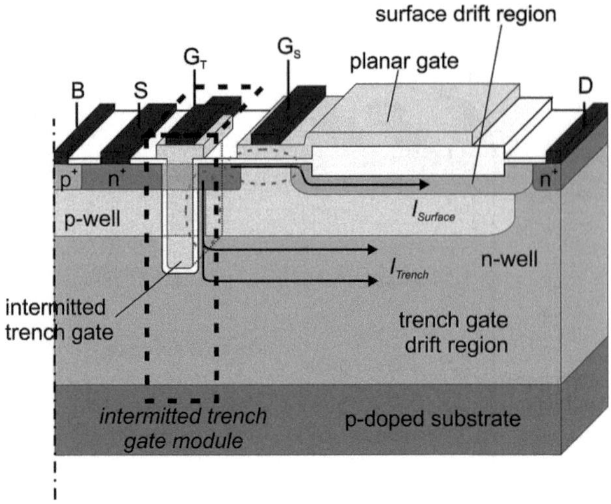

Fig. 8.3 Intermitted trench gate integrated LDMOS transistor (half-cell)—additional trench gate module is indicated by *dashed lines*

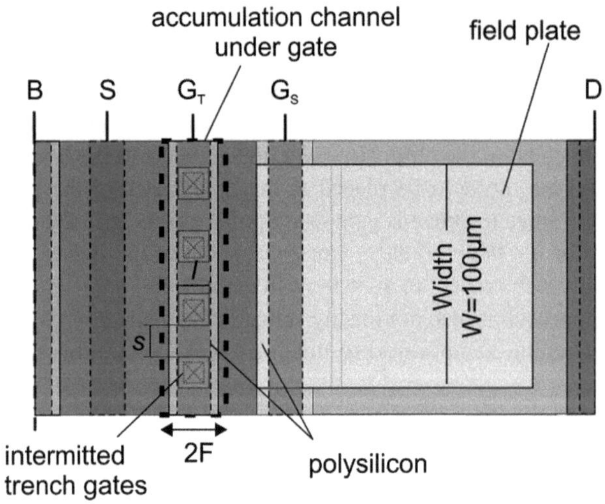

Fig. 8.4 Top view of intermitted trench gate integrated LDMOS transistor (half-cell)—additional trench gate module is indicated by *dashed lines*

This is in contrast to the designs with continuous trench gates where the trench completely intersects the source region. That requires the use of vias to both sides of the trench in order to bias the separated source islands. In the intermitted trench gate design, the source regions below the trench gate electrode cannot easily be doped by ion implantation due to the presence of a continuous polysilicon layer which is used for the self-alignment of the gate when forming the source and drain regions. The doping concentration in the n-well is significantly lower than the source contact region, and this would result in an undesirably high source resistance. But, the formation of an accumulation channel during the on-state of the trench gate region—similar to the accumulation region in VDMOS transistors [9]—leads to a negligible source resistance compared to the resistances of the drift regions. The increase in gate-source capacitance is considerably small.

Moreover, the availability of intermitted trench gates offers the option to adopt the ratio of trench gate to source regions (*l-s* ratio) within the trench gate area in correlation to the drift resistance of the LDMOS device. As discussed in Chap. 7, the channel resistance of LDMOS devices becomes smaller compared to the drift resistance with increasing blocking voltage. Hence, the implementation of a gate with full channel width (identical to the width of the cell) is no longer mandatory. Depending on the trench gate-to-source ratio, this opens up possibilities like placement of the source contacts in line with the intermitted trench gates. Note that for an *s-l* ratio of 1:1, the effective channel width of the intermitted trench gate concept equals to the width of a continuous trench gate, because the channel regions that are perpendicular to the width of the device are also electrically active. A comparison of the contact layout of this intermitted trench gate integrated LDMOS transistor compared to an LDMOS transistor with continuous trench gate is depicted in Fig. 8.5.

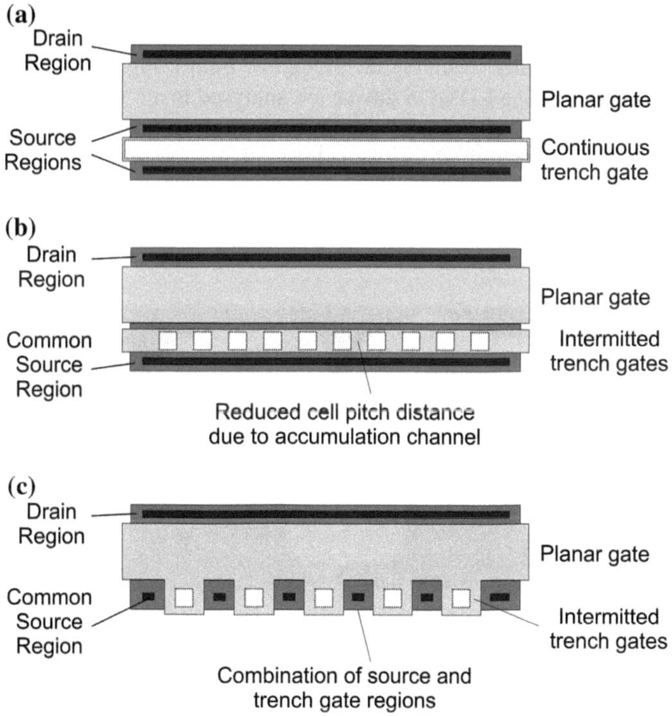

Fig. 8.5 Comparison of contact layout of **a** Continuous, **b** and **c** Intermitted trench gate integrated LDMOS transistor

8.1.3 Verification of Manufacturability

To verify that the trench gate integration concept can be implemented using standard CMOS technology, LDMOS devices without RESURF design have been fabricated using a 2 μm CMOS technology as a proof-of-concept. A cross-section SEM image of a 100 V continuous trench gate integrated LDMOS transistors is presented in Fig. 8.6.

Both planar and trench gate devices were fabricated using the same fabrication process with an additional trench gate module for the latter devices. The gate oxide was 50 nm thick to allow for a gate voltage of 20 V. Doping of the polysilicon gates was performed by diffusion of phosphorous. Therefore, the trench gate width has been increased to allow for conformal deposition of a phosphorus silicate glass on the polysilicon inside the trench. This gate formation technology can be implemented when in situ doping of polysilicon during trench refill is not available. The output characteristics of this continuous trench gate integrated LDMOS transistor are displayed in Fig. 8.7.

It is evident that the drain current through the trench gate and the planar gate can be independently controlled. Also note that operation of both gates simultaneously does not yield the sum of the individual drain currents because they share a common drift region in this simple design.

In order to further investigate the feasibility and limitations of these integration concepts in commercially available technologies, both a high frequency LDMOS device and a high voltage LDMOS device are analysed in a 350 nm smart-power IC process environment. The group of switches for high frequency operation is represented by a 20 V LDMOS with extended drain but without a RESURF topology. Whereas a 50 V LDMOS with RESURF design acts as the demonstrator towards the group of high voltage switches.

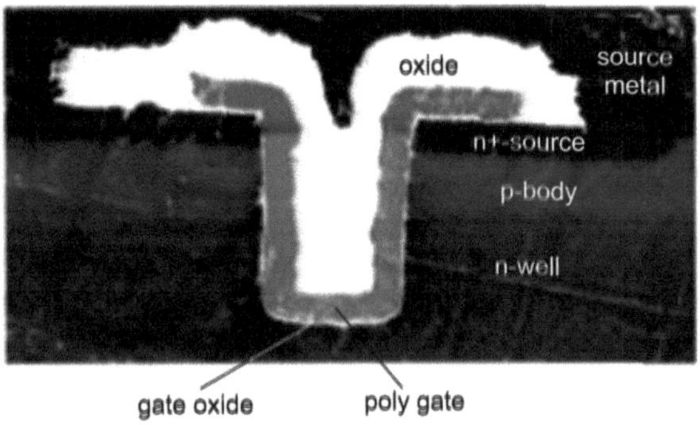

Fig. 8.6 Scanning spreading resistance measurement of trench gate integrated LDMOS transistor in 2 μm technology (gate region is shown)

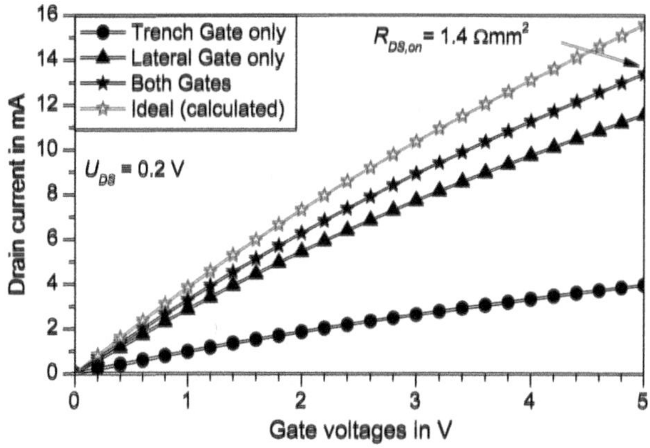

Fig. 8.7 Output characteristics of 140 V CTGI LDMOS device in 2 μm CMOS technology

8.2 Electrical Properties Under Static and Dynamic Device Operation

To judge the impact of trench gate integration on the electrical properties, the on-state characteristics of both *continuous trench gate integrated* (CTGI-) LDMOS transistors and *intermitted trench gate integrated* (ITGI-) LDMOS transistors are investigated first. The following results are extracted from two-dimensional TCAD simulations based on a commercially available Smart-Power IC fabrication process in 350 nm technology.

8.2.1 On-State Resistance

In order to evaluate the reduction of on-state resistance for the CTGI- and ITGI-LDMOS transistors in various design variations with device widths of 100 μm, their transfer characteristics were extracted at a low drain-source voltage of 0.2 V. First, 20 and 50 V CTGI-LDMOS transistors with a gate oxide module of 3.3 V are investigated, which represents the standard gate oxide of the CMOS devices in this technology. This means that a maximum voltage of 3.3 V can be reliably applied to the gate electrode. Unless otherwise noted, a trench depth of 2 μm reaching through the 1.3 μm deep p-well is implemented. The corresponding transfer characteristics are given in Fig. 8.8.

The feasibility of trench gate integration into a planar LDMOS process becomes clearly evident from this figure. In addition to the current through the planar gate, the current through the trench gate contributes significantly to the overall drain current. Especially, the RESURF-based 50 V devices benefit from the trench gate integration where a current increase of almost 100 % is predicted. The $R_{DS,on}$ can be lowered

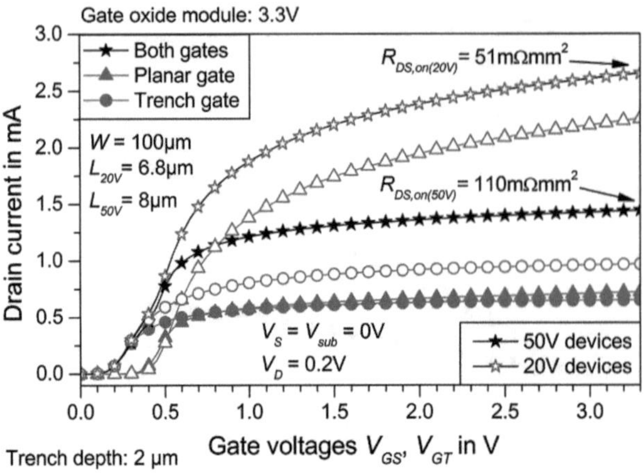

Fig. 8.8 Transfer characteristics of area-neutral continuous trench gate integrated LDMOS transistors for 20 and 50 V operating voltage with a 3.3 V gate oxide module

from 217 mΩ mm^2 for the underlying LDMOS to 110 mΩ mm^2 for the area-neutral CTGI LDMOS with a half-cell length of 8 µm. This represents an $R_{DS,on}$ reduction of 49 %, and demonstrates the advantage of using the n-well of an LDMOS transistor for current conduction. Considering that both the planar drift region and the n-well are capable of blocking the full drain-source voltage by design, the amount of charge in these regions are similar due to the requirements of a working RESURF design. Of course, making use of the whole trench drift region results in significant current spreading and raises its effective length. For the larger half-cell length of the conventional CTGI LDMOS design an $R_{DS,on}$ value of 153 mΩ mm^2 is calculated using a similar approach, which still constitutes to a reduction of 29 %.

The same observations hold true for the 20 V devices (no RESURF) where a reduction from 62 to 51 mΩ mm^2 is observed for the area-neutral design with continuous trench gates (see Fig. 8.2) and a half-cell length of 6.8 µm. However, when the conventional CTGI LDMOS design from Fig. 8.1 is considered, an increase in $R_{DS,on}$ by 3 % to 64 mΩ mm^2 is observed. This means that for this 20 V device, the increase in drain current is less prominent than the increase in length of the half-cell due to integration of the trench gate.

Taking a closer look at Fig. 8.8, it is evident that the threshold voltages between planar and trench gate are asymmetric. The difference between the threshold voltages is due to the graded concentration profile of the p-well along the vertical trench gate in comparison to the constant profile along the device surface for the lateral planar gate. The p-well is fabricated by ion implantation and diffusion, and the doping concentration reduces from the surface towards the junction with the n-well. In addition, the doping concentration is more severely affected by boron segregation during thermal oxidation of the gate oxide formation due to the lower initial doping concentration in the trench region. Consequently, the threshold voltage shifts towards more negative values for the trench gate. This effect further

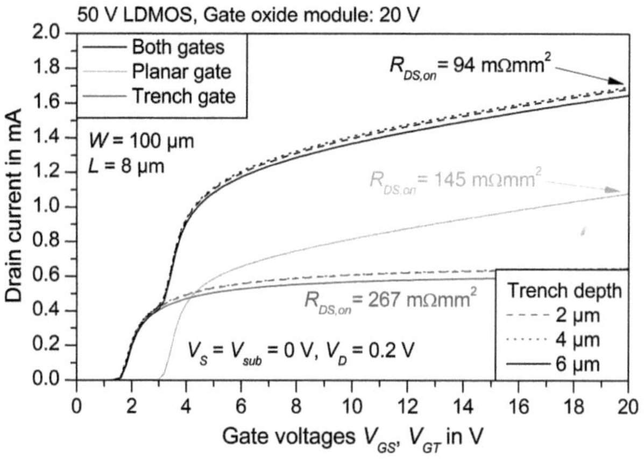

Fig. 8.9 Transfer characteristics of 50 V area-neutral continuous trench gate integrated LDMOS transistors with 20 V gate oxide module for different trench depths

increases when thicker gate oxides are used, which is confirmed from the transfer characteristics of a 50 V LDMOS device with a 20 V gate oxide module shown in Fig. 8.9.

Here, the threshold voltage difference between the planar and trench gate is more than 1 V. In RF LDMOS transistors, this could cause a strong non-linearity when operating in class AB, B or C amplifiers. Nevertheless, this is acceptable for Smart-Power ICs where the LDMOS is primarily used as a digital switch. Here, the leakage current through the trench gate at 0 V gate bias is less than 10 nA. Moreover, a subthreshold swing similar to the planar device is observed.

Figure 8.9 also reveals that the on-state resistance of the LDMOS devices with a 20 V gate oxide module can be reduced. The underlying 50 V LDMOS device exhibits an $R_{DS,on}$ value of 145 mΩ mm^2. Neglecting source, drain and contact resistances, the trench gate constitutes to a parallel conduction path with an $R_{DS,on}$ value of 267 mΩ mm^2. These two parallel resistances result in a combined on-state resistance of 94 mΩ mm^2 which is 36 % lower than operating the planar gate alone. For comparison, the conventional CTGI LDMOS transistor with an increased footprint achieves an $R_{DS,on}$ value of 117 mΩ mm^2.

It can also be deduced that increasing the depth of the gate trench from 2 to 6 µm for the 1.3 µm deep p-well slightly increases the drain current. This observation represents the current spreading effect along the trench. It occurs due to the formation of an accumulation channel in the n-well at the oxide/semiconductor interface at the trench gate and represents an extension of the trench channel similar to the accumulation channel in VDMOS power transistors. The trench current can then pass through the drift region at a lower local current density, i.e. a deeper region of the 6.2 µm deep n-well can be efficiently used for current conduction. However, using a deep trench is strongly discouraged because the benefit is insignificant compared to the vast increase in transfer capacitances contributing to dynamic switching losses (compare Fig. 8.14).

To further reduce the footprint of the LDMOS transistor while maintaining a separate p^+-contact, the intermitted trench gate integration concept was developed. Its advantages are evident from the transfer characteristics shown in Fig. 8.10.

The drain current through the trench gate of the ITGI device with a separate p^+-contact and a half-cell length of 8.7 μm is even higher than for the corresponding CTGI device with a half-cell length of 9.5 μm. The omission of the source contact between the trench and planar gates leads to a reduction of the trench drift region length and thus to a lower drift resistance in the trench drift region. Moreover, the reduction of the device footprint (half-cell length) further lowers the on-state resistance. For the present case, an $R_{DS,on}$ value of 94 mΩ mm^2 is obtained [8]. This value is similar to the on-state resistance of the area-neutral 50 V CTGI LDMOS transistor, but the body of the transistor can be biased separately, e.g. for ultra-low standby power application.

A summary of on-state resistance for different CTGI (area neutral and extended half-cell) devices with an operation voltage of 20 V is given in Table 8.1. Due to the significance of the channel resistance and the prospect of high frequency operation, the ITGI devices are not discussed for this voltage class.

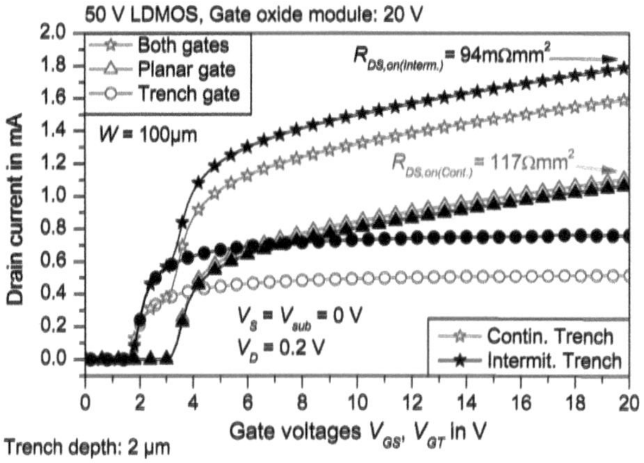

Fig. 8.10 Transfer characteristics of intermitted and continuous trench gate integrated LDMOS transistors with 50 V operating voltage

Table 8.1 On-state resistances of 20 V CTGI LDMOS transistors for different gate oxide modules

20 V devices	Gate oxide module		
	3.3 V	5 V	20 V
Standard LDMOS	62 mΩmm^2	63 mΩmm^2	58 mΩmm^2
Extended CTGI-LDMOS	64 mΩmm^2 (+3 %)	66 mΩmm^2 (+5 %)	59 mΩmm^2 (+2 %)
Area-neutral CTGI-LDMOS	51 mΩmm^2 (−21 %)	53 mΩmm^2 (−16 %)	48 mΩmm^2 (−17 %)

Table 8.2 On-state resistances of 50 V CTGI and ITGI LDMOS transistors for different gate oxide modules

50 V devices	Gate oxide module		
	3.3 V	5 V	20 V
Standard LDMOS	217 mΩmm²	209 mΩmm²	145 mΩmm²
Extended CTGI-LDMOS	153 mΩmm² (−29 %)	151 mΩmm² (−28 %)	117 mΩmm² (−18 %)
Area-neutral CTGI-LDMOS	110 mΩmm² (−49 %)	110 mΩmm² (−47 %)	94 mΩmm² (−35 %)
Extended ITGI-LDMOS	114 mΩmm² (−47 %)	114 mΩmm² (−45 %)	94 mΩmm² (−35 %)

For all gate oxide modules, the increase of the drain current through the additional conduction path along the trench does not make up for the increased cell length of the extended CTGI-LDMOS design. This confirms the assumption that for short drift regions (low blocking voltages) an extended device design is not feasible.

For LDMOS devices with higher operating voltages, the effort of trench gate integration becomes more promising. In Table 8.2 the on-state resistances of 50 V CTGI and ITGI LDMOS transistors are presented.

In summary, for the LDMOS devices with 50 V operating voltage, trench gate integration provides significant $R_{DS,on}$ reduction across all gate oxide modules. The ITGI concept offers values of on-state resistance as low as the area-neutral CTGI concept with the advantage of a separate p-body contact.

8.2.2 Breakdown Voltage

To evaluate the benefit of on-state resistance reduction by integration of trench gates, the breakdown voltages of the 20 and 50 V devices are also investigated. Device failure at excess forward blocking voltages occurs due to avalanche breakdown. As a first order approximation, the breakdown voltage can be evaluated from TCAD simulations where the impact ionization integral equals to unity [10].

Given that no modifications to the wells were performed for trench gate integration (other than etching the trenches), blocking operation is expected to yield similar results for devices with and without trench gate integration. This assumption is confirmed by the forward blocking characteristics presented in Fig. 8.11.

For the 20 V devices, the trench gate higher blocking voltages than the underlying LDMOS device provides. The 20 V devices without RESURF regions significantly benefit from the trench region which acts as a vertical field plate.

In contrast, such a significant increase of breakdown voltage is not observed for the 50 V CTGI and ITGI devices because these devices already feature a RESURF design. The trench gate still acts as a vertical field plate as demonstrated in Fig. 8.12.

For both trench depths, the voltage at the drain voltage is repelled from the surface, limiting the electric field in the trenches to values below 3 MV/cm. This effect is more prominent for the trench with 4 μm depth which extends further into

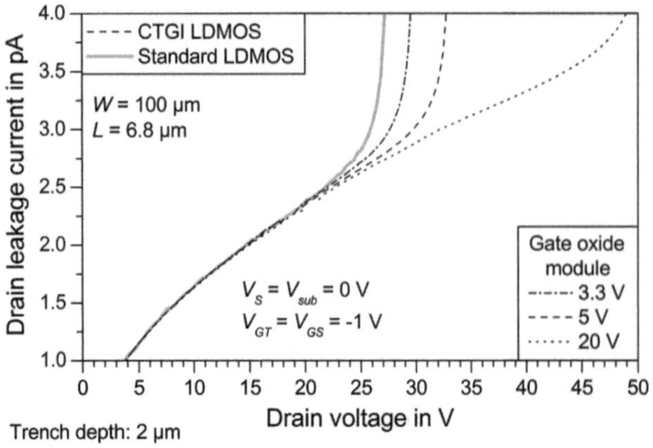

Fig. 8.11 Forward blocking characteristics of continuous trench gate integrated LDMOS transistors with 20 V operating voltage

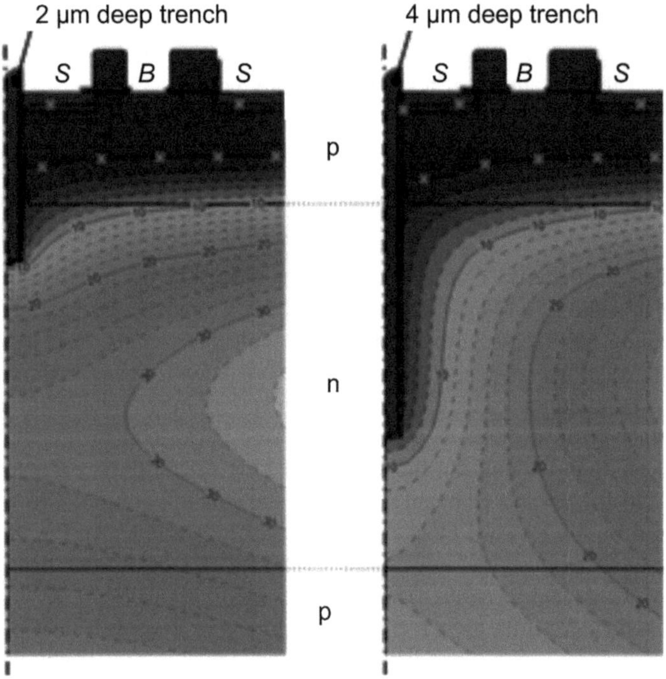

Fig. 8.12 Distribution of electrostatic potential beneath the source regions of 50 V CTGI LDMOS device for 2 and 4 µm deep gate trenches illustrating the vertical field plate effect

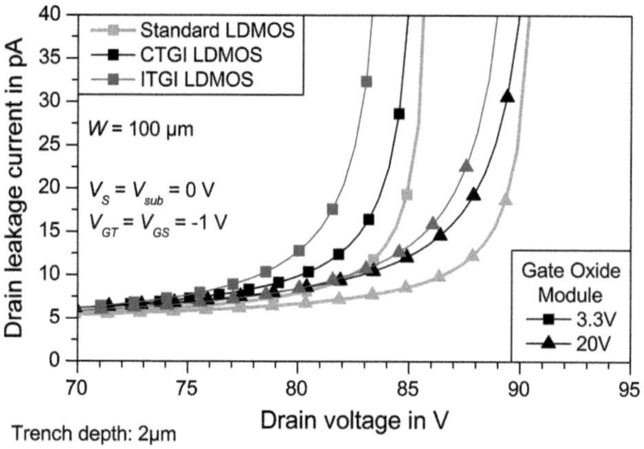

Fig. 8.13 Forward blocking characteristics of intermitted and continuous trench gate integrated LDMOS transistors with 50 V operating voltage and different gate oxide modules

the n-well. But it is also visible for the 2 µm deep trench. The forward blocking characteristics of the 50 V devices are shown in Fig. 8.13.

In this case, the additional charges in the trench gate region constitute to a charge misbalance reducing breakdown voltage compared to the standard LDMOS design. However, the amount of charges is small compared to the overall balancing charge of the RESURF region, and it is confined to the gate region. Hence, the effect is not very prominent. Nevertheless, care should be taken to verify the impact of the charge imbalance when integrating trench gate into an existing LDMOS design.

For both 20 and 50 V devices, an increase in breakdown voltage is more pronounced for thicker gate oxides (higher gate oxide module). This effect can be explained by a larger voltage drop across the oxide below the field plate for a thicker gate dielectric.

The breakdown voltage of the 50 V devices can be extracted from Figs. 8.11 and 8.13 by taking the onset of avalanche coinciding with the asymptotical increase in drain current and the point where the impact ionisation integral becomes unity. These breakdown voltages are presented in Table 8.3. From the previous results, the comparison of on-state resistance versus breakdown voltage can be deducted as summarized in Fig. 8.14.

The fully isolated trench gate integrated LDMOS transistors exhibit on-state resistance similar to non-insulated state-of-the-art RESURF transistors. Due to the

Table 8.3 Breakdown voltage of 50 V integrated trench gate LDMOS transistors for different gate oxide modules

50 V devices	Gate oxide module		
	3.3 V	5 V	20 V
Standard LDMOS (V)	86.3	87.8	91.0
Extended CTGI-LDMOS (V)	85.7	87.5	91.5
Extended ITGI-LDMOS (V)	84.5	86.1	90.6

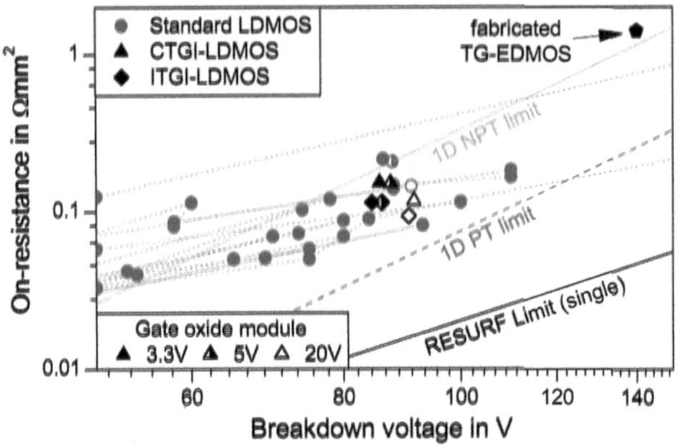

Fig. 8.14 Figure-of-merit for static power dissipation: Trade-off between on-state resistance and breakdown voltage for different 50 V LDMOS devices with respect to the silicon limit

reduction of the on-state resistance, trench gate integrated LDMOS devices exhibit lower static power losses for a given chip area than the planar LDMOS devices. This gives rise to either a more efficient switching circuit or a reduction of overall chip size.

Compared to the superjunction LDMOS transistors introduced in Chap. 6 the trench gate integrated LDMOS devices exhibit lower device resistances for 20 and 50 V devices. This is particularly noteworthy due to the ease of integration of the trench gate integrated LDMOS devices that goes—in the present case—without the necessity to alter any implantations.

8.2.3 Avalanche Ruggedness

Besides the ability of blocking a drain-source voltage in the off-state of the device, another important property is avalanche ruggedness. During switching operation of the LDMOS, both the full blocking voltage is present and a significant forward current occurs at the onset of the Miller-Plateau. Under these dynamic conditions, the high electron density constituting to the drain current represents additional charges in the drift region, and the effective doping concentration is increased. Moreover, these charges also cause a charge imbalance deteriorating the RESURF design [11]. Both effects alter the electric field distribution away from the optimum design resulting in a reduced breakdown voltage. To estimate the immunity against this Kirk effect [12] on the trench gate integrated LDMOS devices, the worst case condition is applied. Due to the saturation in LDMOS devices, the highest drain current occurs for high drain-source voltages at maximum gate voltage. The breakdown voltage at maximum gate voltage is therefore a good estimation for the avalanche ruggedness of the devices. Because additional displacement charge can

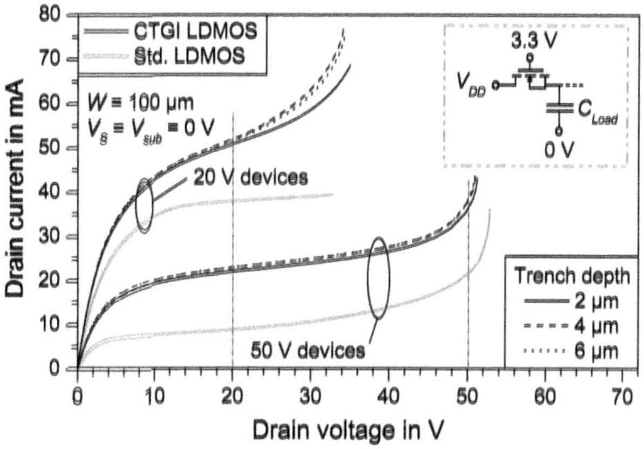

Fig. 8.15 Output characteristics of continuous trench gate integrated LDMOS transistors (20 and 50 V) at maximum gate voltage for avalanche ruggedness analysis

originate from steep voltage ramps in the LDMOS device, the dynamic avalanche should occur well above the maximum rated drain-source voltage.

For the CTGI design, dynamic avalanche was investigated and the corresponding output characteristics are shown in Fig. 8.15 [13].

A significant reduction of breakdown voltage compared to the static off-state cases shown in Figs. 8.11 and 8.13 is evident. The current saturation of the 20 V CTGI devices is overlapped by the onset of impact ionization at drain voltages above 20 V. Generally, the CTGI devices which achieve higher on-state currents show a reduced breakdown voltage for dynamic avalanche as more carriers are available for impact ionization. Nevertheless, the 20 V devices without RESURF design are quite stable with respect to dynamic avalanche effects. These devices exhibit relatively high doping concentrations that would require higher device currents for deterioration to occur.

For the 50 V devices, a significant reduction of breakdown voltage close to 50 V is observed. Here, the additional charge carriers significantly alter the charge distribution in the drift region, reducing the effectiveness of the RESURF design. Again, a higher current density results in a more severe deterioration. This can also be concluded from the dependence of the trench depth on the output characteristics. The breakdown voltage decreases with higher trench depth, even though the increase in drain current for deeper gate trenches is small compared to the total current. In order to account for process fluctuations, it may be necessary to increase the drift region length for the 50 V CTGI LDMOS devices. This would reduce the electric field across the RESURF region and increase the breakdown voltage. Also, it will not have an impact on other devices on the chip because the change only affects the layout of the trench gate integrated devices. The minimum breakdown voltages due to dynamic avalanche for the CTGI LDMOS devices in comparison to the standard LDMOS transistors are summarized in Table 8.4.

Table 8.4 Minimum breakdown voltage due to dynamic avalanche for different CTGI LDMOS devices

20 V devices	Gate oxide module		
	3.3 V	5 V	20 V
Standard LDMOS (V)	24.9	32.2	27.4
Extended CTGI-LDMOS (V)	28.1	>35	34.9
Area-neutral CTGI-LDMOS (V)	29.5	>35	34.0

50 V devices	Gate oxide module		
	3.3 V	5 V	20 V
Standard LDMOS (V)	53.2	53.2	52.8
Extended CTGI-LDMOS (V)	50.8	52.3	50.8
Extended ITGI-LDMOS (V)	50.7	52.8	50.8

8.2.4 Drawbacks Under Operation as High-Side Switches

The trench gate integration concepts were introduced based on fully-isolated LDMOS transistor cells which are capable of blocking the drain voltage against the wafer backside. This is enabled by a junction isolation featuring a RESURF design for the n-well as previously shown, e.g. in Fig. 8.1. The isolation to the substrate is a prerequisite for the implementation of these LDMOS transistors in monolithic integrated buck converters or in half and full bridge configurations. Here, the transistor highlighted in Fig. 8.16 acts as a high-side switch with its source potential ranging from 0 V to supply voltage level and the drain potential being fixed to supply voltage level.

Without a suitable isolation of the high-side switch, a short circuit condition would occur during operation of this circuitry. In order to prevent this, the substrate is continuously biased to 0 V and the elevated source potential is isolated by an additional n-well which forms a blocking pn-junction to the p-doped substrate [4].

For trench gate integrated LDMOS devices which make use of the n-well for current conduction, this can cause an increase in on-state resistance when operating as a high-side switch. The reason is the extension of the space-charge region

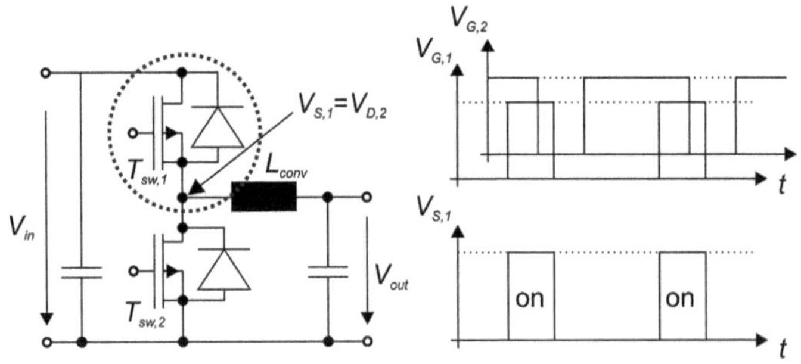

Fig. 8.16 Half bridge circuit topology including an LDMOS transistor as a high-side switch

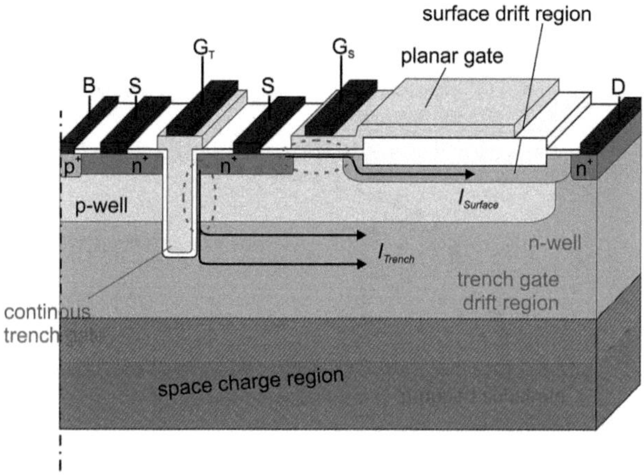

Fig. 8.17 Reduction of trench gate drift region width by presence of a space-charge region in a CTGI LDMOS transistor operated as a high-side switch

occurring across the substrate/n-well junction. This effect is depicted in Fig. 8.17 for the CTGI LDMOS design from Fig. 8.1, and it holds true for the other TGI designs as well.

Subsequently, the effective depth of the trench drift region available for current conduction is reduced. The resulting transfer characteristics of a 50 V CTGI LDMOS device in forward conduction when operated as a high-side switch in comparison to a low-side switch can be deduced from Fig. 8.18.

The results confirm that the resistance of the trench drift region is increased. For the area-neutral CTGI device, the on-state resistance increases from 94 to 108 mΩ mm^2. This value is still 25 % lower than the $R_{DS,on}$ value of the standard LDMOS device. As the high-side switch of the half-bridge depicted in Fig. 8.16

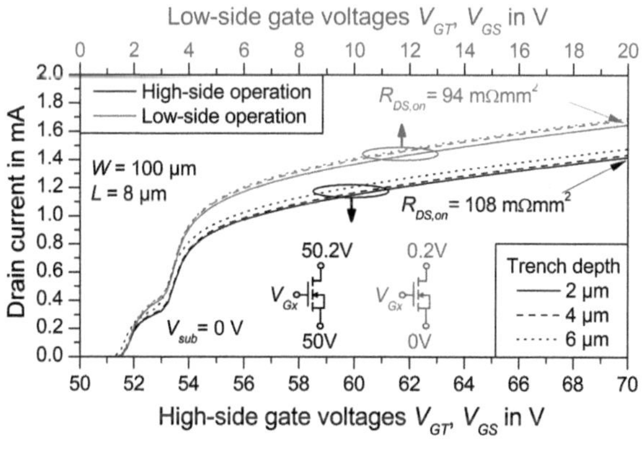

Fig. 8.18 Transfer characteristics of 50 V area-neutral CTGI LDMOS transistor (20 V gate oxide module) operating as a high-side switch compared to characteristics for low-side operation

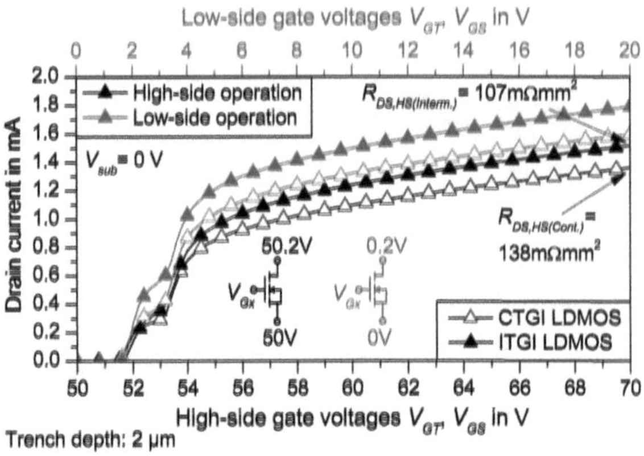

Fig. 8.19 Transfer characteristics of 50 V extended CTGI and ITGI LDMOS transistors (20 V gate oxide module) operating as a high-side switch compared to characteristics for low-side operation

will not conduct current in steady-state when the source potential is zero (switch 1 open and switch 2 closed), the difference in drift region resistance will not be of significance for most applications. Nonetheless, care should be taken during circuit design to account for the increase of $R_{DS,on}$ under high-side operating conditions. Also note that the impact of trench depth is insignificant for the operation as a high-side switch. Similar considerations hold true for the extended CTGI and ITGT concepts. Their transfer characteristics under high-side and low-side operation are presented in Fig. 8.19 for comparison.

Again, the ITGI device experiences an increase of $R_{DS,on}$ from 94 to 107 mΩ mm^2. However, the on-state resistance of the extended CTGI device becomes 138 mΩ mm^2. This is close to the value of the standard LDMOS (145 mΩ mm^2) with the smaller cell footprint. Therefore, the extended CTGI device is not an attractive choice for implementation as a high-side switch.

8.2.5 Switching Losses Under High Power Operation

A key factor for energy conversion circuits in smart-power ICs is energy efficiency. Typically, it determines the device area of the LDMOS transistors, i.e. the number of cells that constitute to one device. Naturally, the device area should be as small as possible to allow for a small overall chip size. However, running high current densities across small areas causes a large amount of power dissipation. Therefore, a trade-off must be considered when designing the device area for an LDMOS transistor. Static power dissipation can be easily determined by the on-state resistance. However, dynamic power dissipation, which occurs during device switching, also needs to be considered. This is particularly true for high switching frequencies encountered in smart-power ICs and/or small duty cycles at low load conditions where low static losses with respect to dynamic losses incur.

Given that the integration of trench gates introduces further gate capacitances, the impact on dynamic power dissipation needs to be considered. A typical turn-on event for an LDMOS device assuming a constant gate charge current and an inductive load was introduced in Chap. 4. The main portion of dynamic power dissipation arises from the charging of the transfer capacitance during phase III (Miller plateau [14]) when the device is on the verge of turning on and a high drain current is already flowing through the device. At the same time, the drain-source voltage is still high and gradually reduces while the transfer capacitance is charged. The amount of this gate-drain charge Q_{gd} required is indicative to the amount of power losses during this dynamic switching event as it determines the duration of the Miller plateau. Moreover, it is independent of the gate current [9].

Integrating the transfer-capacitance from supply voltage down to the drain-source voltage for a given current density yields a decent approximation of the gate-drain charge Q_{gd} for investigation of the different trench gate integration concepts. For this purpose, a supply voltage of 50 V and a drain current at which the Miller voltage equals to 3 V were selected. The calculated gate charges for the different integration concepts are presented in Table 8.5.

As expected, the device concepts employing additional trench gates are characterized by a higher gate charge, i.e. more charge is required to turn the devices on and off. Conversely, for a given gate resistance, a higher gate charge means that switching takes longer and higher dynamic power dissipation takes place.

Still, it should be considered that the trench and planar gates act on the same p-well. The contribution of the planar gate to the gate charge from Table 8.5 (values shown in brackets) implies that this vertical field plate effect also reduces the gate-drain capacitance of the planar gate. The feasibility of trench gate integration into planar LDMOS topology with respect to dynamic switching losses is judged using the $R_{DS,on} \cdot Q_{GD}$ Figure-of-Merit shown in Fig. 8.20.

The results show that the increased gate charge is well compensated by the lower on-state resistance, making the CTGI and ITGI devices promising candidates for energy efficient power devices on a reduced chip area.

Table 8.5 Comparison of gate charges during switching of CTGI and ITGI LDMOS devices for different gate oxide modules; values in brackets indicate the contribution of the planar gate

50 V devices	Gate oxide module		
	3.3 V	5 V	20 V
Standard LDMOS (pC/mm^2)	584	624	882
Extended CTGI-LDMOS (pC/mm^2)	729 (417)	747 (422)	1,133 (732)
Extended ITGI-LDMOS (pC/mm^2)	877 (439)	842 (445)	1,810 (805)

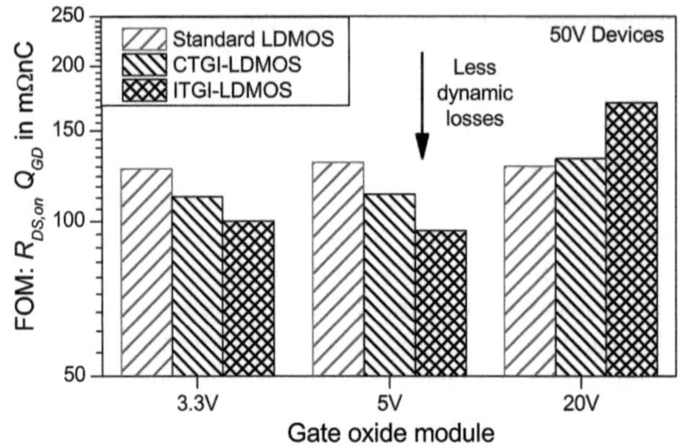

Fig. 8.20 Figure-of-Merit $R_{DS,on} \cdot Q_{GD}$ for dynamic power dissipation for CTGI and ITGI LDMOS devices

8.2.6 Switching Losses Under High Frequency Operation

The evaluation of the gate charge is a first indication that high frequency operation using the CTGI or ITGI LDMOS devices is restricted due to the increased device capacitances. To further analyse the suitability of these devices regarding radio frequency operation, the input capacitances (gate-source capacitance) are extracted from small-signal simulations. Based on the gate-source voltage, the input capacitance determines the amount of charge that is consumed during each device turn-on cycle (and dissipated during each turn-off cycle). For high switching frequencies, charging of the input capacitance resembles a major source of power losses. Moreover, a high input capacitance limits the switching frequency simply by demanding very large gate currents, and consequently large gate drivers. The input capacitances for the different 50 V ITG LDMOS devices are summarized in Table 8.6.

All trench gate integrated device design exhibit a significantly higher input capacitance for a 2 μm deep trench. While there is still some room for improvement by reducing the trench depth (straight-forward) and the depth of the channel and source wells (requires LDMOS redesign), it is apparent that the trench gate induces high values of input capacitance because it acts between gate and source. This is illustrated in Fig. 8.21.

Table 8.6 Comparison of input capacitances of CTGI and ITGI LDMOS devices for different gate oxide modules

50 V devices	Gate oxide module		
	3.3 V	5 V	20 V
Standard LDMOS (pF/mm^2)	830	400	130
Extended CTGI-LDMOS (pF/mm^2)	1,780	750	160
Extended ITGI-LDMOS (pF/mm^2)	3,480	1,540	390

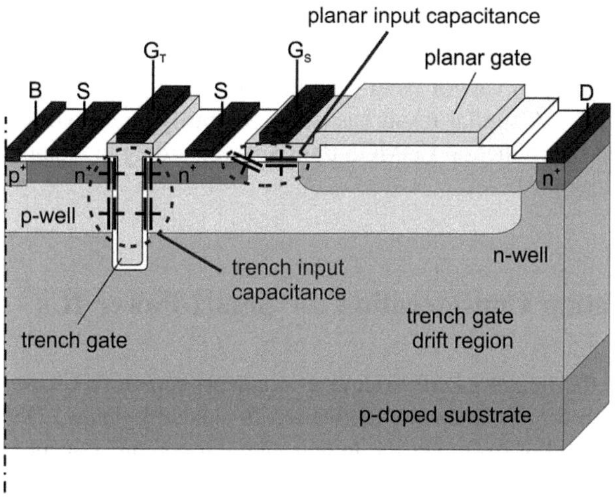

Fig. 8.21 Contributions to input capacitance in extended CTGI LDMOS device from Fig. 8.1

Since the p-body has to be fully intersected by the trench gate to ensure an inversion channel through the p-well, a shallower trench will reduce the transfer capacitance but not the input capacitance. Still, for the ITGI design a less pronounced increase in input capacitance may be achieved when reducing the trench gate-to-source region ratio. Moreover, the increase in input capacitance is smaller for thicker gate oxides. This is due to the smaller gate oxide capacitance which is connected in series to the substrate capacitances in the source and p-body regions next to the trench gate. Additionally, the vertical doping of the p-well is also reducing towards the bottom of the well, locally reducing the substrate capacitance as well.

To estimate the impact of trench gate integration on the power losses for high frequency operation, the $R_{DS,on} \cdot C_{in}$ Figure-of-Merit can be used [9]. The results for extended ITGI and CTGI devices are displayed in Fig. 8.22.

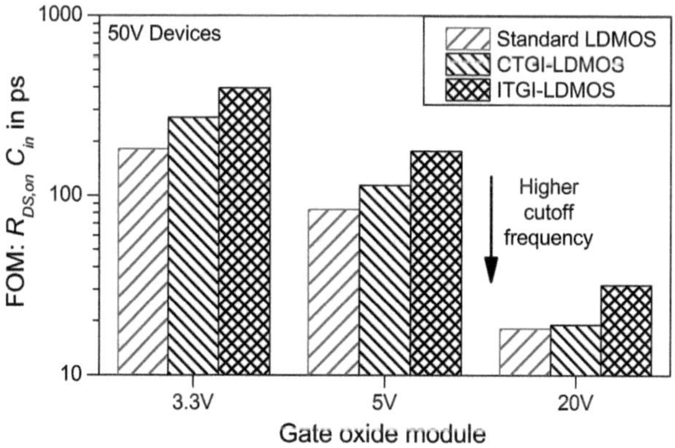

Fig. 8.22 Figure-of-Merit $R_{DS,on} \cdot C_{in}$ for power dissipation at high operating frequencies for CTGI and ITGI LDMOS devices

This Figure-of-Merit is a direct indication of the minimum power losses that can be achieved in the device. It can be concluded that the combination of planar and trench gate topologies cannot readily be used for high frequency operation in the upper MHz range. A similar result has been previously published for lateral trench gate integrated superjunction LDMOS transistors regarding their applicability in RF LDMOS devices [15].

8.3 Integration Consideration for Smart-Power ICs

In contrast to the trench LDMOS device concepts shown in Chap. 7, this device topology allows for simultaneous fabrication of standard planar LDMOS and TGI-LDMOS devices. For example, the trench gate integration concept can be applied for the 50 V LDMOS devices while using the standard 20 V LDMOS devices for a different voltage level converter on the same chip.

After implantation of wells, trench gates are patterned by reactive ion etching as shown in Chap. 7. The trench gates are oxidized and doped polysilicon is deposited which allows for a simultaneous gate stack formation of the trench and planar channel region. Note that the deposition of in situ doped polysilicon and completely refilling the trenches yields the easiest integration approach. In the case that ex situ doping of the polysilicon by low pressure chemical vapour deposition (LPCVD) using $POCl_3$ and a drive-in or by subsequent ion implantation is required, additional effort is necessary to form the trench gates. One option is to increase the trench width to prevent a refill of the trench by the polysilicon deposition. Then, the polysilicon can be doped by tilted implants or by a LPCVD process. Given that polysilicon thickness is typically in the range of 250–500 nm (e.g. [16]), this does not constitute to a significant increase in cell pitch. In this case, the trenches will be refilled by the oxide passivation.

A significant reduction of on-state resistance by up to 49 % has been predicted in 50 V LDMOS transistors using intermitted trench gates. For this commercially available 350 nm smart-power IC technology, the wells of the underlying LDMOS devices are suited for a "More-than-Moore" integration approach. Sufficient ruggedness against dynamic avalanche conditions occurring during switching events has been demonstrated in general. Due to the increased forward current densities, care must be taken to avoid occurrence of avalanche under all dynamic switching conditions. When operated as high-side device, the trench gate integrated LDMOS suffers from the extension of the space-charge region from the p-substrate into the n-well, but the increase in $R_{DS,on}$ is still well below that of the standard 50 V LDMOS. When operating at high frequencies, the high input capacitance of the additional trench gate negatively impacts device performance. For operating frequencies typically encountered in integrated switch mode converters, a clear reduction of static and dynamic power dissipation can be expected from integration of trench gates into the planar LDMOS topology.

References

1. D. Disney, W. Chan, R. Lam, R. Blattner, S. Ma, W. Seng, J.-W. Chen, M. Cornell, R. Williams, 60 V Lateral Trench MOSFET in 0.35 μm technology, in *International Symposium on Power Semiconductor Devices and ICs*, Orlando, USA, 2008
2. I.-Y. Park, C.A.T. Salama, CMOS compatible super junction LDMOST with N-Buffer layer, in *International Symposium on Power Semiconductor Devices and ICs*, Santa Barbara, CA 2005
3. T. Erlbacher, A.J. Bauer, L. Frey, Reduced on resistance in LDMOS devices by integrating trench gates into planar technology. IEEE Electron Device Lett. **34**, 464–466 2010
4. B. Murari, F. Bertotti, G.A. Vignola, *Smart Power ICs*. (Berlin, Springer, 2002)
5. M. Bergmann, *An Introduction to Many-Valued and Fuzzy Logic: Semantics, Algebras, and Derivation Systems, New York* (Cambridge University Press, USA, 2008)
6. C.-C. Cheng, H.L. Chou, F.Y. Chu, R.S. Liou, Y.C. Lin, K.M. Wu, Y.C. Jong, C.L. Tsai, J. Cai, H.C. Tuan, Investigation of parasitic BJT turn-on enhanced two-stage drain saturation current in high-voltage NLDMOS. in *International Symposium on Power Semiconductor Devices & ICs*, San Diego, USA 2011
7. F. Fallah, M. Pedram, Standby and active leakage current control and minimization in CMOS VLSI Circuits. IEICI Trans. **88**-C(4), 509–519 (2005)
8. T. Erlbacher, A. J. Bauer, L. Frey, Significant on-resistance reduction of LDMOS devices by intermitted trench gates integration. IEEE Trans. Electron Devices **59**, 3470–3476 (2012)
9. B.J. Baliga, *Fundamentals of Power Semiconductor Devices, New York* (Springer, USA, 2008)
10. R. van Overstraeten, H. de Man, Measurement of the ionizsation rates of diffused silicon p-n junctions. Solid-State Electr. **13**, 583–608 (1970)
11. A.W. Ludikhuize, Kirk Effect Limitations in High Voltage IC's, in *International Symposium on Power Semiconductor Devices and ICs*, Davos, Switzerland 1994
12. C.T. Kirk, A theory of transistor cutoff frequency (fT) falloff at high current densities. IRE Trans. Electron Devices ED **9**(2), 164–174 (1962)
13. T. Erlbacher, G. Rattmann, A.J. Bauer, L. Frey, Trench gate integration into planar technology for reduced on-resistane in LDMOS devices, in *International Symposium on Power Semiconductor Devices and ICs*, Hiroshima, Japan, 2010
14. J.M. Miller, Dependence of the input impedance of a three-electrode vacuum tube upon the load in the plate circuit. Sci. Papers Bur. Stan. **15**(351), 367–385 (1920)
15. I. Cortes, P. Fernandez-Martinez, D. Flores, S. Hidalgo, J. Rebollo, Analysis of low-voltage super-junction LDMOS structures on thin-SOI substrates. Semicond. Sci. Technol. **23**, 015009 (2008)
16. M.J. Kumar, R. Sithanandam, Extended-p+ stepped gate LDMOS for improved performance. IEEE Trans. Electron Devices **57**(7), 1719–1724 (2010)

Chapter 9
Lateral Power Transistors on Wide Bandgap Semiconductors

For radio frequency applications, lateral GaAs heterojunction bipolar transistors (HBT) are routinely employed in the RF amplifier of mobile phones, and silicon RF LDMOS transistors are used in the corresponding base station equipment. Due to the bandgap engineering performed on the GaAs transistors, they offer a significantly higher transition frequency than silicon transistors. However, output power density in GaAs based RF amplifiers is limited due to low breakdown voltage in spite of very high electron mobility [1]. Over the last few decades, a different class of lateral power transistors has evolved using wide bandgap semiconductors on silicon carbide and gallium nitride which will be discussed in the course of this chapter. An overview on physical and electrical properties of silicon and 4H-silicon carbide with importance regarding RF and power electronic applications is shown in Fig. 9.1.

The high critical breakdown field has driven the development of power semiconductor devices on silicon carbide and gallium nitride for power electronic applications. Motivation for using wide bandgap semiconductors stems from the benefit of high operating voltages for high output power which allows for an output impedance close to 50 Ω according to (3.3). This minimizes the effort for matching networks [2, 3].

In this chapter, the state-of-the-art and progress on lateral silicon carbide power transistors of the 4H polytype is introduced. The impact of the physical properties on electrical characteristics is explained and the benefits regarding power electronic and radio-frequency applications are discussed. Technological limitations like limited channel mobility due to high interface state densities will be considered. Both work towards high frequency and high voltage operation are considered. Aspects regarding the introduction of smart power ICs on silicon carbide like the state-of-the-art in high temperature SiC logic will be investigated.

Moreover, the high electron mobility transistor (HEMT) on gallium nitride is in the focus of this chapter. Due to the formation of a two dimensional electron gas it is highly suitable for high frequency operation. The performance of GaN-based HEMTs in its current state is described. Then, reliability issues like dynamic on-resistance and current collapse which hamper introduction of these devices in power electronic circuits are explained. Requirements regarding the epitaxial layers for fabrication of HEMTs and the available quality for GaN-on-Silicon devices are

T. Erlbacher, *Lateral Power Transistors in Integrated Circuits*,
Power Systems, DOI 10.1007/978-3-319-00500-3_9

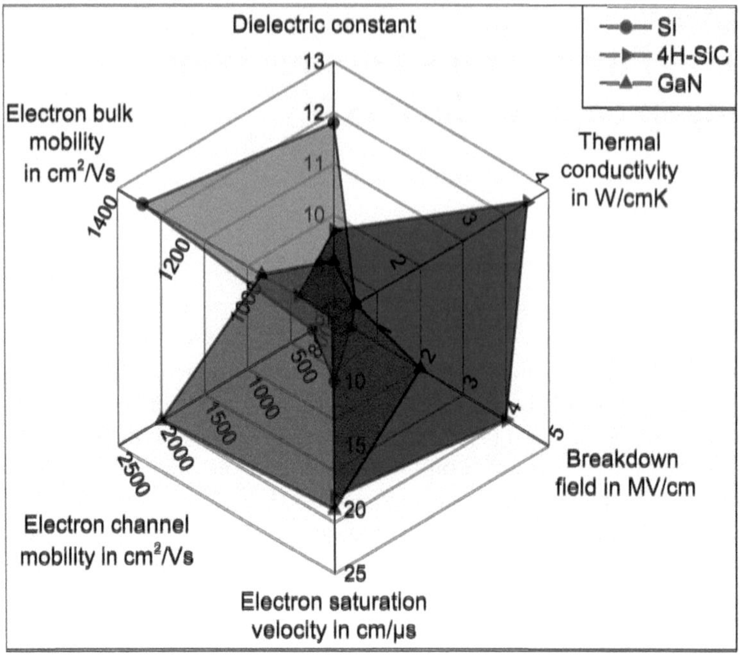

Fig. 9.1 Physical and electrical properties of silicon, and silicon carbide (4H polytype) relevant for power semiconductor devices

discussed. Using this technology, considerations regarding the implementation of GaN HEMTs on local GaN islands on silicon-based CMOS circuits for implementation of smart-power ICs and RF power amplifiers are presented.

9.1 Lateral Silicon Carbide Transistors on 4H Polytype

Over the last decade significant development towards silicon carbide power semiconductor devices was conducted. Device technology for vertical power transistors on the 4H-polytype of silicon carbide has arrived at the production stage [4]. A stable and reliable processing technology is vital to the development of devices for power electronic and RF applications. Therefore, in the course of this textbook only the 4H polytype of silicon carbide is considered.

9.1.1 Material Properties of 4H Silicon Carbide

However, developments on lateral silicon carbide transistors are scarce despite several material properties making 4H-SiC superior to silicon devices—at least on the paper.

9.1.1.1 Critical Electric Field and Drift Region Resistance

The bandgap of 3.2 eV for 4H-SiC requires higher electric fields for impact ionisation of free charge carriers to occur. The impact ionisation coefficient for electron and holes in 4H-SiC is significantly higher than in silicon. For analytical evaluation, approximations for the impact ionisation coefficients have been derived by Fulop [5] for silicon and by Baliga [6] for 4H-SiC. Using these approximations and solving the ionisation integral, the critical electric field for 4H-SiC is in the range of 2–4 MV/cm depending on the doping concentration. As depicted in Fig. 9.2, it is almost one decade larger than the critical electric field for silicon.

Considering the larger extension of the space-charge region (due to the higher voltage), the breakdown voltage of a 4H-SiC pn-junction is up to 56 times higher than a silicon pn-junction with similar doping concentration.

Consequently, the drift region resistance of power semiconductor devices for a given breakdown voltage is also vastly reduced. This is the key driver for the development of SiC technology as of today. Recalling (5.8), the drift region resistance of a RESURF based LDMOS transistor is given by

$$R_{drift,RESURF} \cdot A \geq \frac{1}{\mu_{n,drift}\varepsilon_0\varepsilon_r E_{crit}^3} \cdot \sqrt{2} \cdot BV^2. \tag{9.1}$$

The breakdown voltage depends on the critical electric field, and the following relationship is derived using Baliga's approximation for 4H-SiC [6]:

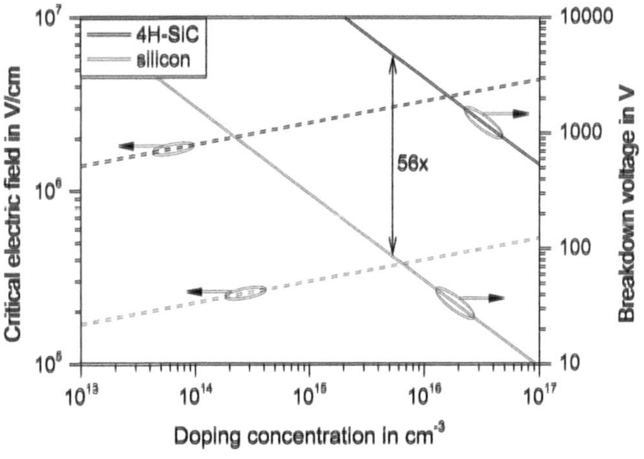

Fig. 9.2 Critical electric fields for silicon and 4H-SiC and breakdown voltages under non punch-through conditions based on the approximations for impact ionisation coefficients after Fulop [5] and Baliga [6]

$$\frac{E_{crit,Si}}{V\,\mathrm{cm}^{-1}} = 1.25 \cdot 10^7 \cdot \left(\frac{BV}{V}\right)^{-\frac{1}{6}}. \tag{9.2}$$

Then, the area-specific drift region resistance limit of a RESURF-LDMOS transistor on 4H-SiC is given by

$$\frac{R_{drift,RESURF} \cdot A}{\Omega\,\mathrm{cm}^2} \geq 1.31 \cdot 10^{-12} \cdot \left(\frac{BV}{V}\right)^{2.5}, \tag{9.3}$$

which is over 1,500 times smaller than the value for silicon RESURF-LDMOS transistors from (5.10). Regarding drift region resistance, the smaller bulk mobility of 4H-SiC is compensated by the high critical electric field.

9.1.1.2 Electron Drift Velocity and Electron Mobility

Regarding high frequency operation, the saturated electron drift velocity has to be considered. As with silicon, drift velocity in 4H-SiC can be approximated by the field dependent equation of a group IV semiconductor.

$$v_{IV-IV}(E) = \frac{\mu E}{\left[1 + \left(\frac{\mu E}{v_s}\right)^{\alpha}\right]^{\frac{1}{\alpha}}}, \tag{9.4}$$

with the electric field E, the saturation drift velocity v_s and the fitting parameter α [7]. For III–V semiconductors like GaAs and GaN, a more complex approximation formula is required to account for the mobility peak before saturation of drift velocity occurs:

$$v_{III-V}(E) = \frac{\mu E + v_s \frac{E^{(n_1-1)}}{E_c^{n_1}} E}{1 + a\left(\frac{\mu E}{v_s}\right)^{n_2} + \left(\frac{\mu E}{v_s}\right)^{n_1}}, \tag{9.5}$$

with the fitting parameters given in [8]. Temperature dependence of the electron velocity on the electric field was also modelled using Monte-Carlo simulations [9]. A comparison of drift velocities for (100) silicon, GaN and (0001) 4H-SiC is shown in Fig. 9.3. The drift velocities were measured using the conductance technique [11]. Silicon carbide offers a significantly higher saturated electron drift velocity than silicon. This is particularly true for high temperatures where the reduction of the saturated electron drift velocity in silicon carbide is less pronounced than in silicon.

It should also be noted that the electron drift velocity in 4H-SiC is lower compared to silicon in the linear regime before saturation occurs. This also results in lower inversion layer mobility in a MOSFET under high vertical electric fields

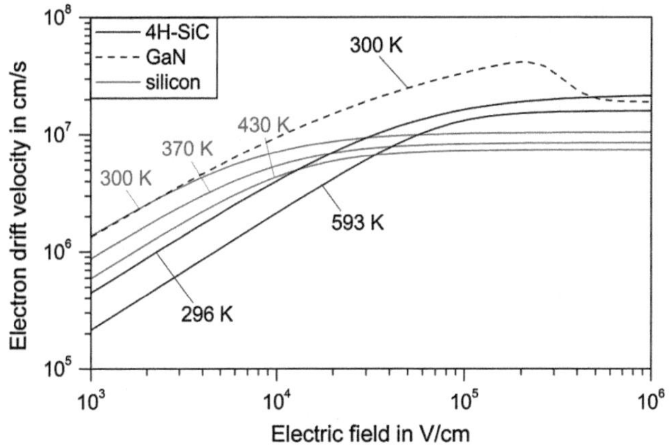

Fig. 9.3 Modelled drift velocities for electrons in silicon, silicon carbide and gallium nitride for different temperatures, fit parameters from [7, 10]

compared to silicon. At an electric field of 1 MV/cm which typically results from the gate voltage applied to a MOSFET, the inversion layer mobility in silicon is in the range of 200–300 cm^2/Vs [12]. In contrast, inversion channel mobility in 4H-SiC MOSFETs is significantly lower and strongly depends on the fabrication processes [13]. Currently, it is technologically possible to fabricate stable MOSFETs with inversion layer mobility of approximately 25 cm^2/Vs on 4H-SiC. The physical reasons and the technological obstacles are discussed in more detail in the course of this chapter.

Considering the discussion on the channel resistance in Chap. 7 and including the critical electric field dependence from (9.3), a minimum area-specific channel resistance for 4H-SiC devices can be estimated similar to (7.7) for silicon LDMOS transistors:

$$R_{Ch,min} \cdot A = 38610 \cdot \frac{F}{\mathrm{cm}} \left(\frac{\left(\frac{BV}{V}\right)^{\frac{7}{6}}}{1.25 \cdot 10^7} + 11 \cdot \frac{F}{\mathrm{cm}} \right). \tag{9.6}$$

In Fig. 9.4 theoretical limits for the drift and channel regions of silicon and 4H-SiC LDMOS transistors are summarized. The effect of lower mobility on drift region and channel resistance has to be considered together with the benefits from the increased critical electric field. But clearly, it constitutes to a technological limitation that is yet to be overcome.

Comparing these drift region and channel limits, the voltage region where silicon carbide exhibits significant advantages becomes apparent. The drift region resistance of 4H-SiC LDMOS transistors plays a significant role for devices with breakdown voltages in excess of 1,000 V when an inversion channel mobility of 250 cm^2/Vs similar to silicon is assumed. For the current state-of-the-art technology (25 cm^2/Vs), the drift region resistance will be negligible for breakdown voltages up to

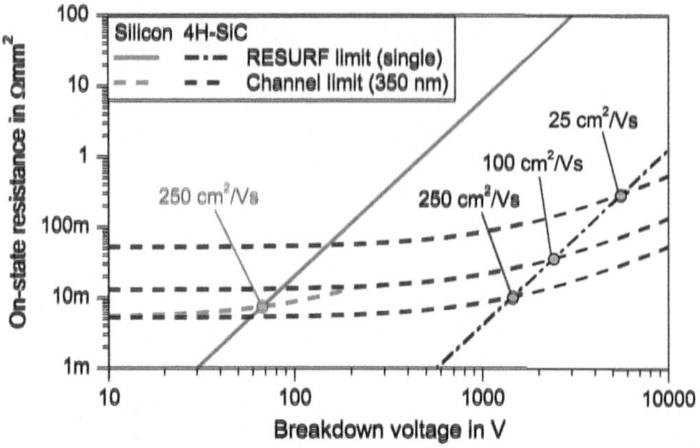

Fig. 9.4 Drift region resistance (RESURF-based) and channel resistance limits for LDMOS transistors fabricated on silicon and 4H-SiC; several values of inversion channel mobility for 4H-SiC are depicted

approximately 3 kV. This value is significantly higher than for vertical SiC devices where a higher channel density can be achieved, e.g. by VDMOS or trench designs. Below these voltages, the resistance of 4H-SiC devices is governed by the channel resistance which exhibits a weak dependence on breakdown voltage. Therefore, for breakdown voltages over 200 V, these devices can achieve lower total device resistance than their silicon counterparts. The research efforts regarding the fabrication of 4H-SiC LDMOS transistors and combination of 4H-SiC LDMOS transistors with logic functionality are discussed in Sects. 9.1.2 and 9.1.4, respectively.

9.1.1.3 Intrinsic Carrier Concentration

Another implication of the wide bandgap is the lower intrinsic carrier concentration compared to silicon. It stems directly from the wide energy bandgap of the semiconductor. The low probability of thermal generation of electron-hole pairs manifests in a negligible amount of minority charge carriers at room temperature for doping concentrations typically encountered in semiconductor devices. For silicon, an intrinsic carrier concentration at room temperature of $1.08 \cdot 10^{10}$ cm^{-3} has been widely accepted [14]. The intrinsic doping concentration for 4H-SiC at room temperature was theoretically determined to be approximately $5 \cdot 10^{-9}$ cm^{-3}, which is over 18 decades smaller than in silicon [15]. Considering the temperature dependence of the density of states and the bandgap, the calculated intrinsic carrier concentration for 4H-SiC are compared to values for silicon and GaN in Fig. 9.5.

This strong difference in intrinsic carrier concentration compared to silicon has severe implications on high temperature operation for both wide bandgap materials. The intrinsic carrier concentration of silicon approaches values of doping concentrations typically encountered in power semiconductor devices at about 500 K.

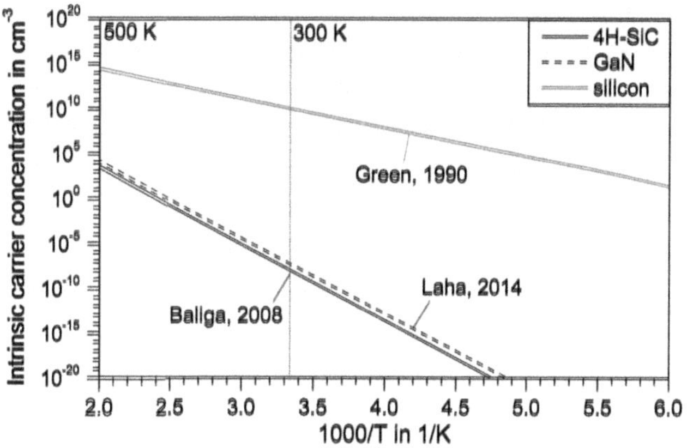

Fig. 9.5 Intrinsic carrier concentration from 165 to 500 K for silicon, 4H-SiC and GaN calculated from [6, 14, 16], respectively

Therefore, device performance will be deteriorated due to the intrinsic conduction for temperatures over 200 °C and barely any silicon device technology can address these harsh operating conditions [17]. Additionally, the generation current of pn-junctions under reverse bias is directly proportional to the intrinsic carrier concentration [18]. Leakage current in power semiconductor devices and CMOS circuitry increases exponentially with temperature and current consumption may already exceed specifications well below 200 °C.

Silicon carbide and gallium nitride devices offer much higher operating temperatures because of the lower intrinsic barrier concentration. Of course, this requires adequate contact metallization and assembly technologies to exploit this benefit. Moreover, leakage currents in pn-junctions on SiC are also significantly lower compared to silicon devices, allowing for high temperature, low leakage applications to be served. Operation of SiC MOSFETs at up to 923 K has been reported [19].

9.1.1.4 Thermal Conductivity

Finally, the high thermal conductivity of silicon carbide substrates should also be emphasized. While there has not been much work performed to this regard, measurements indicate that the thermal conductivity of 4H-SiC at 300 K is approximately 3.7 W/cmK [20]. This is over two times higher than the room temperature value of 1.5 W/cmK for silicon [21].

Considering Fourier's Law of heat conduction, this implies that a higher power density can be used in an assembly (e.g. with a package and a heat sink) than for a silicon device. A detailed analysis of thermal resistance in MOSFETs depending on bias conditions was given by Wang et al [22]. A simplified thermal circuit of a power semiconductor device mounted on a heat sink is given in Fig. 9.6.

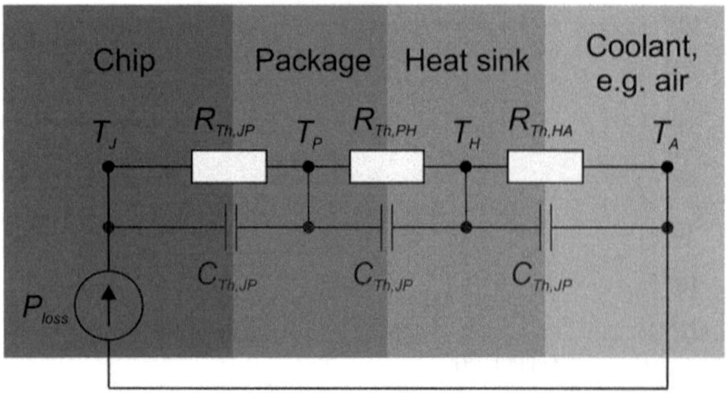

Fig. 9.6 Thermal equivalent circuit for a packaged semiconductor device mounted to a heat sink

The thermal resistance of the silicon or silicon carbide material is modelled using $R_{TH,\ JP}$. This value is—in a first order approximation—directly proportional to the thermal conductivity of the semiconductor substrate. Silicon carbide devices can be operated at higher power densities without increasing junction temperature. The achievable increase in power density strongly depends on the thermal conductivity of the package, the heat sink and the coolant. Combining this observation with the capability of 4H-SiC to operate at higher junction temperatures than silicon for the reasons mentioned in the previous section, higher power densities and elevated ambient temperatures can be used than in silicon.

9.1.2 Progress on Silicon Carbide MeSFETs and MOSFETs

With the aforementioned properties, silicon carbide is an attractive material both for power electronic and radio-frequency applications. Similar to silicon technology, different device topologies have evolved to address the demands of these two markets. Even more, different device concepts like MeSFETs, JFETs and MOSFET based power transistors are still competing to find both an entrance into existing silicon-dominated domains and to enable new systems with better-than-ever performance.

9.1.2.1 Devices Tailored Towards Radio Frequency Applications

Due to the high saturation velocity and the high breakdown voltage, 4H-SiC transistors for RF applications have been investigated to great extent. This development has been spearheaded by designs of metal-semiconductor field effect transistors (MeSFETs), i.e. metal electrodes without a gate dielectric in direct contact to the semiconductor forming a Schottky barrier. A typical MeSFET design on a semi-insulating substrate is illustrated in Fig. 9.7.

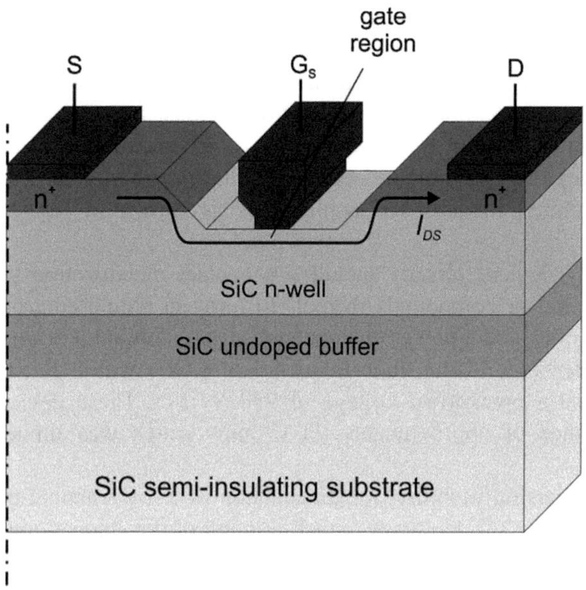

Fig. 9.7 RF MeSFET on semi-insulating 4H-SiC substrate, adopted from [23]

The device current is controlled by applying a negative voltage to the gate which leads to an extension of the space-charge region from the surface towards the semi-insulating substrate. Thereby, the conductive part of the n-well is reduced and the device can be turned off. Due to the high saturation velocity of electrons which is twice as high as in silicon or gallium arsenide, higher maximum drain currents can be produced. Combined with a higher breakdown voltage, SiC MeSFETs can achieve power densities which are over one decade higher than in silicon or gallium arsenide [24].

Fabrication of these devices is straight-forward and does not require gate oxidation. Limitations to the inversion mobility could be avoided, which was a major issue when development of lateral RF transistors on 4H-SiC started. The leakage current of the reverse-biased gate-substrate Schottky barrier is negligible at room temperature. High transition frequencies require the implementation of short gate regions (ideally in the sub-micron range), which limits the blocking capability of these MeSFETs in silicon carbide to 100–200 V.

For RF MeSFETs on semi-insulating 4H-SiC with a breakdown voltage of 100 V that can be operated at approximately 40–50 V, the transition frequency has been improved from 14 up to 24 GHz within one decade [25–27].

Similarly, first reports on devices with a breakdown voltage of 140 V achieved a transition frequency of 4 GHz and a maximum oscillation frequency of 9 GHz using a 400 nm long gate [28]. Further improvement has been demonstrated by the additional implementation of a faraday shield (source field plate) to minimize the gate-drain capacitance of the MeSFET and the reduction of gate length. The maximum oscillation frequency of this type of devices achieved up to 16 GHz [29]. Finally, 4H-SiC RF MeSFETs with a breakdown voltage of 200 V using a gate length of 1.2 μm and a transition frequency of 20 GHz have also been fabricated [30].

In parallel, investigations on RF LDMOS transistors on 4H-SiC have also received attention despite the limitations in inversion mobility in an attempt to achieve RF amplifiers with higher operating voltages. Initially, basic extended-drain MOSFETs tailored towards RF applications as shown in Fig. 9.8 have been investigated.

These early devices already included a contact metallization (from titanium, aluminum, nickel or compounds thereof) to form an ohmic contact to the highly doped SiC regions and a power metallization (from aluminum or gold) to carry the high current densities. A transition frequency of 8 GHz with a gate length of 1 μm was realized at a breakdown voltage of 950 V [31]. These devices exhibited a device resistance of approximately 24 Ω mm^2 which was limited by channel resistance.

For high inversion mobility, implementation of a low channel doping concentration is desirable [32]. However, punch-through of the channel region could then occur for high drain-source voltage. A buried p$^+$-doped well (anti-punch-through layer), which is implemented using a retrograde doping profile in the channel region helps to overcome these contradicting requirements by providing a low surface doping concentration. Hereby, punch-through is prevented due to the high doping concentration in the buried p-well region. The ohmic contact to this buried p$^+$-layer can be implemented by additional implantation of a p$^+$-sinker or by forming a recess. Moreover, the electric field at the gate-drain overlap can be reduced because this portion of the drift region will be depleted from the buried p$^+$-well. Degradation due to hot carrier injection can be reduced, which is a severe issue due to the higher permissible electric field in silicon carbide compared to silicon. The design of an RF LDMOS with an additional p$^+$-well acting as anti-punch-through layer is schematically depicted in Fig. 9.9.

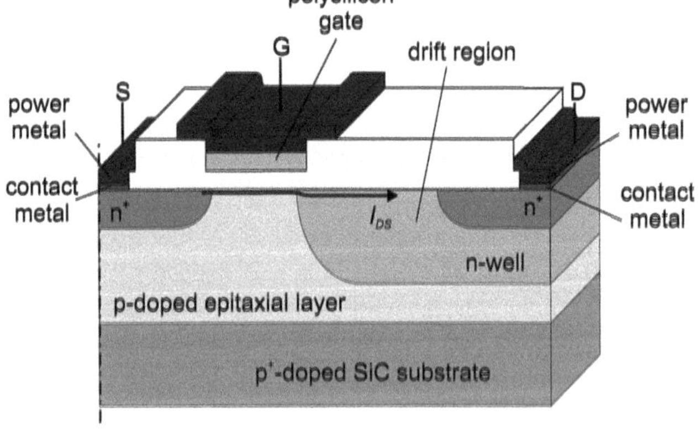

Fig. 9.8 RF LDMOS on 4H-SiC with ohmic contact metallization, adopted from [31]

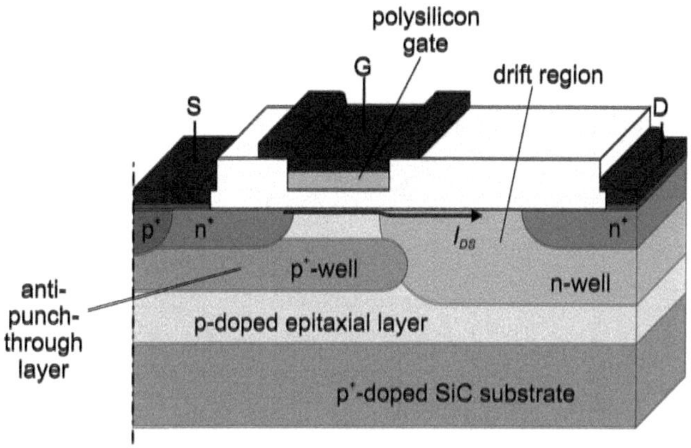

Fig. 9.9 RF LDMOS on 4H-SiC with additional anti-punch-through layer, adopted from [33]

Measured electrical performance of fabricated 4H-SiC LDMOS transistors with a drain extension and the anti-punch-through layer indicates a minimum on-state resistance of 0.16 Ω mm^2 and a breakdown voltage of 90 V. These devices achieved a transition frequency of 11.2 GHz at a gate length of 0.5 μm. A transition frequency of 8.0 GHz was measured for a device with a breakdown voltage in excess of 200 V and a drift region resistance of 0.18 Ω mm^2 [2].

Electrical properties of 4H-SiC devices tailored towards RF applications collected from literature are given in Table 9.1.

A judgment regarding applicability of 4H-SiC RF transistors can be derived from the gain-bandwidth product. As this value also depends on the circuitry (e.g. amplifier class), the maximum oscillation frequency offers a first indication of the RF performance. A comparison of maximum oscillation frequency for different breakdown voltages is presented in Fig. 9.10.

Silicon carbide MeSFET transistors achieve excellent RF performance with breakdown voltages in excess of 100 V. PAE of 57 % for an class AB RF amplifier with a power gain of 15.7 dB and an output power of 4 W/mm were measured for a MeSFET transistor operated at 3.5 GHz and biased at 48 V.

Table 9.1 Breakdown voltage, transition and maximum oscillation frequencies for RF MeSFET and LDMOS transistors fabricated on 4H-SiC

BV (V)	f_T (GHz)	f_{max} (GHz)	Device type	Substrate	Source
100	14	30.5	RF MeSFET	4H-SiC	[25]
140	4	9	RF MeSFET	4H-SiC	[28]
120	–	15.7	RF MeSFET	4H-SiC	[29]
200	8	11.9	RF LDMOS	4H-SiC	[33]
950	7	7.8	RF LDMOS	4H-SiC	[31]
45.3	8.6	11.1	RF MeSFET	4H-SiC	[34]

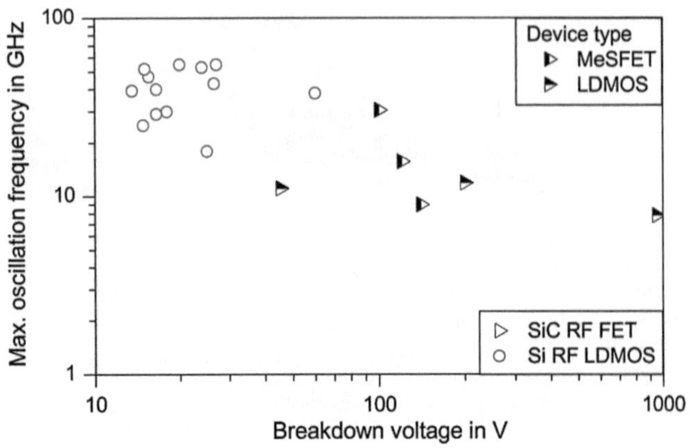

Fig. 9.10 Breakdown voltage versus maximum oscillation frequency for 4H-SiC MeSFETs and LDMOS transistors compared to silicon state-of-the-art RF devices, data from Table 9.1

The 4H-SiC LDMOS transistors also exhibit adequate RF performance even for operating voltages of 50 V. An RF LDMOS operated in class A mode at 3 GHz achieved a PAE of 17.5 % with a power gain of 9.1 dB and an output power of 32 dBm (equals 1.9 W/mm) [2].

9.1.2.2 Devices Tailored Towards Power Electronic Applications

In order to exploit the high voltage capability of silicon carbide in lateral devices, LDMOS transistors have been developed by several groups. As device concepts on silicon carbide do not significantly differ from silicon, the RESURF concept has been adopted early on. The topology for the first LDMOS transistor fabricated on 4H-SiC is schematically illustrated in Fig. 9.11.

These first devices achieved breakdown voltages in excess of 2.6 kV but suffered from poor inversion mobility. A device resistance of 55 kΩ mm^2 was measured due to inversion mobility of less than 10^{-2} cm^2/Vs [35]. Additionally, in order to protect the gate dielectric from the high electric field which silicon carbide brings along, devices with gate oxide thickness of up to 900 μm were first employed [36]. This gave rise to undesirably high threshold voltages, but allowed for fabrication of first SiC devices with a breakdown voltage of 1,200 V and a device resistance of 400 Ω mm^2 which is close to the silicon RESURF limit of LDMOS transistors. Further improvements by reduction of RESURF dose allowed for application of 200 nm thick gate oxides and a field plate design as depicted in Fig. 9.12.

With this concept, the driving capability of these transistors is enhanced, and LDMOS transistors with breakdown voltage of 900 V and device resistance of 50 Ω mm^2 were realized with an inversion mobility of 10 cm^2/Vs. Unfortunately, they still required a gate-source voltage of 80 V [37]. Also, these LDMOS devices have been fabricated on p-doped or semi-insulating SiC substrates.

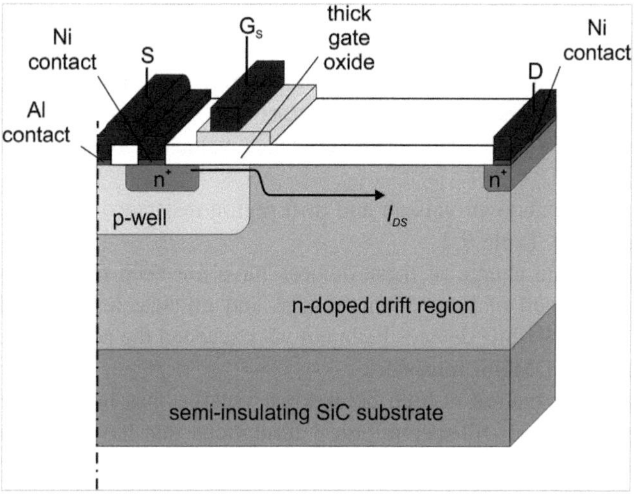

Fig. 9.11 RESURF-based LDMOS transistor on 4H-SiC with thick gate oxide, adopted from [35]

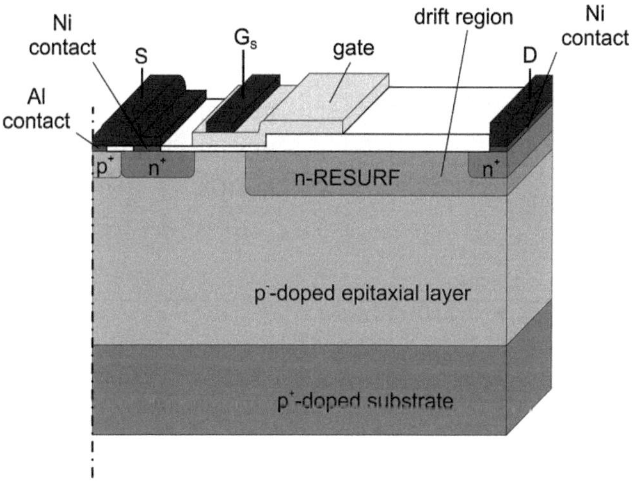

Fig. 9.12 RESURF-based LDMOS transistor on 4H-SiC with reduced RESURF dose and thin gate oxide, adopted from [37]

Haeublein et al. have recently published results on LDMOS transistors fabricated on n^+-doped 4H-SiC substrates with extended drain on n-doped substrates operated under 20 V gate bias that achieved a blocking voltage of 165 V and an on-state resistance of 1 Ω mm^2 [38]. These devices still suffer from the low inversion mobility and the gate length of 2 μm that were used in the study. It was observed, however, that the device resistance decreased with temperature. At 300 °C a reduction of on-state resistance to 0.75 Ω mm^2 can be estimated from the measurement data. A threshold voltage shift, increased inversion mobility or a complete ionisation of dopants at elevated temperatures may be responsible for this observation. RESURF-based LDMOS transistors are expected to achieve similar (channel resistance limited) on-state resistance values at breakdown voltages in excess of 1,000 V.

The data on breakdown voltage and drift region resistance reported in literature are summarized in Table 9.2.

Data on the gate charge of these devices have not been reported. It is evident from the comparison of breakdown voltage and on-state resistance displayed in Fig. 9.13 that the 4H-SiC devices have not yet exceeded the performance of silicon RESURF-based LDMOS transistors.

Even though operation at high breakdown voltages has been demonstrated for SiC devices, there are still technological limitations that have to be overcome to turn the promise of high-voltage low loss SiC LDMOS transistors into reality.

Table 9.2 Breakdown voltage, on-state resistance and gate charge measured from LDMOS transistors fabricated on 4H-SiC reported in literature

BV (V)	$R_{DS,on}$ (Ω mm^2)	Q_{gd} (nC/mm^2)	Device type	Substrate	Source
950	24		EDMOS	4H-SiC	[31]
2,600	55,000		EDMOS	4H-SiC	[35]
1,200	400		RESURF LDMOS	4H-SiC	[36]
900	50		RESURF LDMOS	4H-SiC	[37]
165	1		EDMOS	4H-SiC	[38]

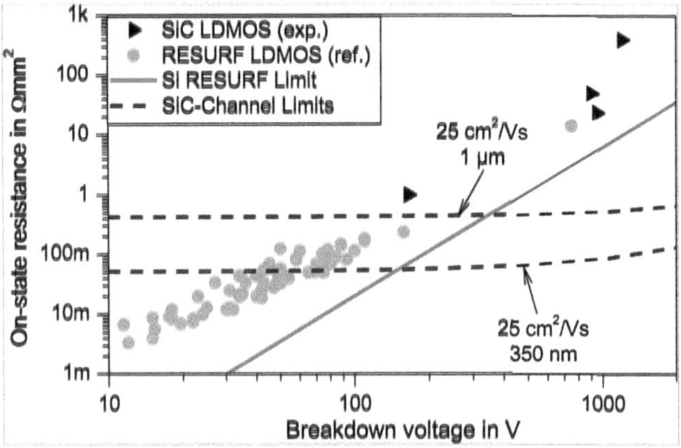

Fig. 9.13 Breakdown voltage versus on-state resistance for SiC LDMOS transistors compared to state-of-the-art silicon devices; silicon RESURF limit and SiC channel limit (for a 350 nm technology) are given for reference, data from Table 9.2

9.1.3 Technological Limitations for 4H-SiC Devices

The development previously described is hampered by insufficient control of processing technology which deteriorates electrical properties and lifetime. Even though silicon carbide devices use similar topologies and methodology as their silicon counterparts, additional processing equipment is required for their fabrication. Primarily, due to the strong Si–C bond, higher temperatures are required for epitaxial growth, activation of dopants after implantation and thermal oxidation.

Semi-Insulating Substrates for RF Devices One of the key constraints for MeSFET-based RF applications is the lack of low defect semi-insulating SiC substrates which are mandatory for the fabrication of large area devices for high power densities with adequate yield. Much of the effort in growing large 4H-SiC single crystals has gone towards power electronics and its demand in highly n-doped material. Therefore, LJFETs and LDMOS transistors are a promising choice as RF power devices as they can also be fabricated on this highly n-doped material. Even though the quality of these substrates is steadily improving and wafers with larger diameter are becoming available, more effort was spent to eliminate micro pipes as the most severe defect type. Especially for operating at high voltages, additional work is being conducted to also minimize dislocations in the wafer or the epitaxy [39]. For example, properly controlling the growth condition of epitaxial layers allow for a conversion of basal plane dislocations (BPDs) into threading edge dislocations [40].

Incomplete Activation of Dopants Another limitation in device operation is the incomplete activation of dopants in 4H-SiC, i.e. the number of dopant atoms that are incorporated into the lattice to provide charge carriers. Much effort has been particularly dedicated to improve the activation of aluminum because its ionisation energy is lower than boron as a p-doping element in 4H-SiC [41]. Similar considerations hold true for nitrogen as the preferred n-doping element instead of phosphorous. Early work on the annealing of ion-implanted dopants was hampered by the silicon-based processing equipment. With the availability of annealing furnaces capable of annealing at temperatures beyond 1,400 °C, it was found that activation of aluminum implanted at the solubility limit increased from 6 to 35 % by increasing the annealing temperature from 1,600 to 1,750 °C. Additionally, implanting the aluminum ions at 1,000 °C prior to annealing allowed for almost 100 % activation under similar annealing conditions [42].

Inversion Mobility While inversion mobility has been significantly improved from values below 0.01 to 20 cm^2/Vs under high gate voltages [32, 35], it is still not close to the theoretical value [43]. The inversion mobility in 4H-SiC MOS-transistors is still limited due to the trapping of free electrons at the interface due to a high density of interface states [32]. This also results in high threshold voltages limiting the inversion charge that is present in the channel from the gate overdrive. Nevertheless, new gate oxidation techniques and the use of lower channel doping concentrations resulted in continuous progress over the recent years, and a final solution to this issue can be expected—especially considering the development that silicon had to undergo to yield oxide-semiconductor interfaces suitable for MOSFET fabrication including chlorine based oxidation and post metallization forming gas annealing [44]. Recently, reduction of interface state density in thin silicon dioxide layers thermally grown on 4H-SiC was observed for an oxidation procedure that was derived from kinetic oxidation analysis [45].

9.1.4 Smart-Power ICs on Silicon Carbide

With the availability of robust lateral power transistors on silicon carbide, smart power ICs like a one-chip power converter with integrated circuit monitoring and protection could be implemented. SiC-on-Si substrates with adequately low defect densities and on large wafer diameters have not been manufactured so far despite early successes on the 3C polytype of SiC [46]. Therefore, homogeneous integration of SiC power devices with SiC logic devices appears as the most promising, yet expensive, approach towards smart-power ICs on silicon carbide. Nevertheless, SiC logic offers operation at temperatures that are out of scope for conventional silicon technology.

In silicon carbide, however, device development has mainly been pushed by the demand of low-loss power electronic switches with high breakdown voltages. The development of a SiC-based logic technology suitable for realization of smart-power ICs is lagging behind in contrast to silicon technology.

Nevertheless, investigations have been carried out towards silicon carbide logic for high temperature applications. These developments started on 6H-SiC substrates which were more readily available at that time, achieving higher inversion mobility and transconductance than their counterparts on 4H-SiC [39]. In 1994, an integrated inverter circuits using n-channel MOS transistors in 6H-SiC was demonstrated [47]. The circuit applied an n-channel MOS driver transistor and a smaller n-channel MOS load transistor. This circuit operated at 304 °C but these transistors were too small to drive any loads and oscillation frequencies could not be investigated.

In 1999, a complete monolithic mixed-signal IC using CMOS technology was presented on 6H-SiC operating at up to 300 °C [48]. This IC implemented an intelligent gate drive circuit which includes a drive circuit and a sensing and protection circuit as depicted in Fig. 9.14.

The drive circuit consists of a gate driver stage that is powered by a charge pump. Additionally, the drive circuit can be disabled in the case of over- or under-voltage detection or by short and open load conditions respectively. All circuit components were fabricated using n-channel MOS and p-channel MOS transistors with distinct size in a circuit design that achieves temperature independent operation. That is particularly important when operating temperature can vary across several hundred Kelvin.

In order to achieve stable high temperature operation over the full lifetime of the devices, metallization and passivation techniques for SiC devices and ICs have also been developed.

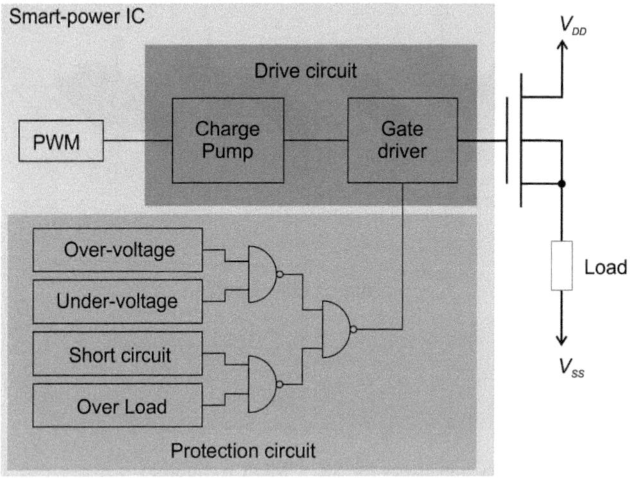

Fig. 9.14 Block diagram of first smart-power IC fabricated in silicon carbide (6H polytype) [48]

For temperatures of up to 500 °C, a platinum metallization with $Ti/TaSi_2$ adhesion layers and a subsequent silicon nitride passivation have been found to be suitable [49].

With respect to 4H-SiC, work on logic transistors started with lateral JFETs which could be used without the need of a gate oxide giving rise to high interface state densities. Switching frequencies in excess of 20 MHz and normally-off power switches operating at up to 900 V using a trench-based lateral JFET technology have been published in 2006 [50]. A similar approach was reported the year after with a 4H-SiC lateral JFET-Based Power IC that contained a vertical channel [51]. The high-voltage devices additionally employed a double-RESURF-based drift region achieving a breakdown voltage of up to 1,028 V.

Analog functions on 4H-SiC have been also demonstrated using MeSFETs. A voltage gain of 16 was obtained from a 200 kHz sine signal from single stage amplifier [52]. A common source topology was implemented for ease of integration.

With respect to logic-level MOS transistors on 4H-SiC, developments took off during the last decade. Using n-channel MOSFETs on 4H-SiC, nMOS logic for operation at up to 400 °C with a TTL compatible operating voltage of 5 V have been fabricated [53]. This technology also employed a high-temperature capable metallization (PtTi with Ti/TiN adhesion layers) and passivation (PECVD-deposited silicon oxide and nitride) technology [54]. Common source amplifiers with stable gain of 6.8 at 300 °C were implemented using depletion mode n-channel MOSFETS [55]. Edge-triggered flip-flops using enhancement-mode 4H-SiC n-MOSFETs operate at frequencies up to 10 kHz at 300 °C and a supply voltage of 15 V [56].

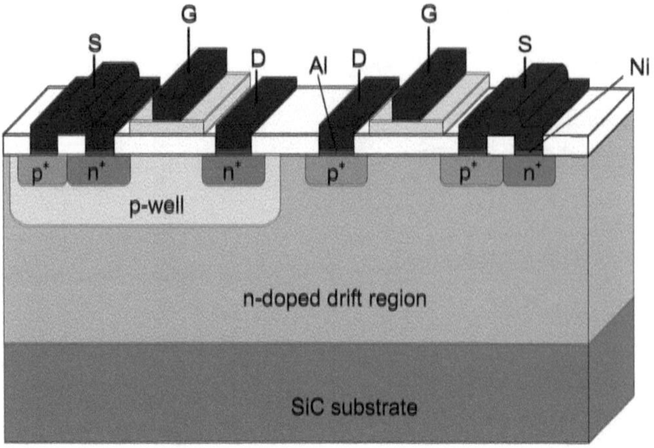

Fig. 9.15 nMOS and pMOS transistors for SiC CMOS technology

For full mixed-signal capability, CMOS technology in 4H-SiC is desirable among other wide bandgap materials because circuit concepts developed in silicon CMOS technology developed over the last decades could be readily transferred. Additionally, fabrication of smart-power ICs in CMOS technology minimizes the effort for integration of lateral MOS-gated power transistors. The cross-section of a SiC-based CMOS technology consisting of an n-channel and a p-channel MOSFET is depicted in Fig. 9.15.

The first CMOS inverter on 4H-SiC, which operated at up to 300 °C with drain-source voltage of 10 V, was presented in 2006 [57]. In 2010, a CMOS-based ring oscillator with 401 stages was constructed that operated at 300 °C and achieved an output frequency of 63 kHz [58].

Silicon carbide HV-CMOS appears as an attractive approach to high temperature applications with the option to design-in RF MeSFET transistors.

9.2 High Electron Mobility Transistors on Gallium Nitride

Transistors on gallium nitride have also drawn great attention over the last decade due to their extraordinary RF performance. A comparison between key physical properties of silicon and gallium nitride is shown in Fig. 9.16.

Since free-standing GaN crystals are not yet readily available, most of the work focused on hetero-epitaxial growth of GaN layers on silicon carbide or sapphire using electrically isolating buffer layers which exhibit lattice mismatch small enough to fabricate GaN layers with defect densities sufficiently low for device fabrication. The work that is underway to grow GaN on large diameter silicon substrates will be assessed later. In either case, the GaN-based devices are fabricated as lateral devices because an undoped GaN buffer layer is required for

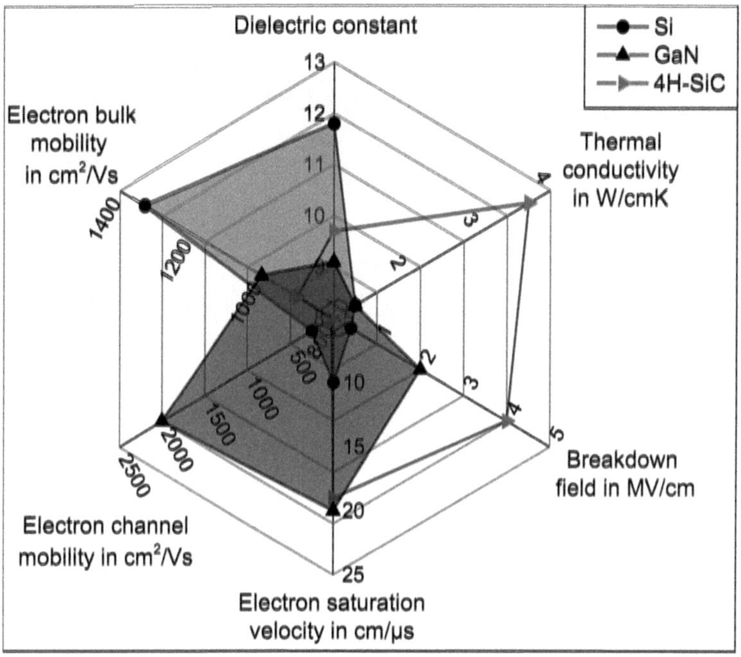

Fig. 9.16 Physical and electrical properties of silicon, and gallium nitride relevant for power semiconductor devices; silicon carbide (4H polytype) from Fig. 9.1 shown for reference

operation of high electron mobility transistors (HEMTs). A cross-section of an AlGaN/GaN HEMT is shown in Fig. 9.17.

This device consists of an undoped GaN buffer and a thin AlGaN layer. A highly conductive electron region is formed at the interface of these two layers. It represents the channel of the transistor. The source and drain electrodes of the transistor are fabricated by etching through the AlGaN layer into the GaN buffer in order to establish an electric contact to this electron channel. The presence of the electron channel without a gate electrode constitutes to the normally-on operation mode typically encountered in AlGaN/GaN HEMTs. Finally, control of the transistor current can be achieved by placing a gate electrode in contact to the AlGaN layer. Then, a negative gate bias can locally reduce the electron concentration in the channel, depleting the electron channel, and turning the transistor off.

9.2.1 Mobility and Two-Dimensional Electron Gas

Gallium nitride exhibits high electron mobility with a distinct mobility peak as shown in Fig. 9.3 before saturation velocity is reached. The behavior is similar to other III–V group semiconductors like GaAs. The combination of spontaneous and

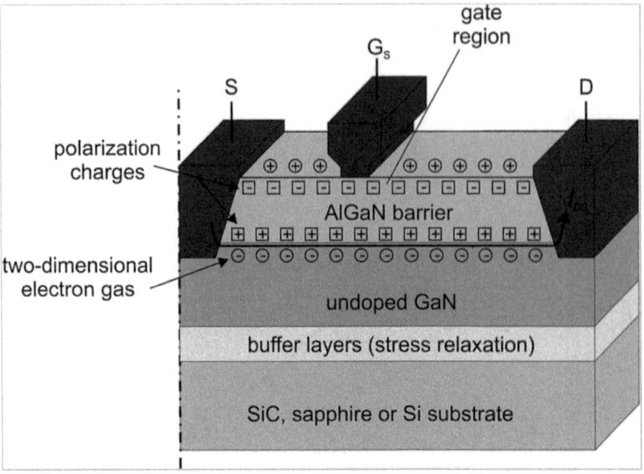

Fig. 9.17 Fundamental AlGaN/GaN high electron mobility transistor topology for RF applications

piezoelectric polarization causes significant band bending at the AlGaN-GaN interface as illustrated in Fig. 9.18.

Electrons confined in this potential well exhibit very high mobility in a very narrow region which is referred to as a quantized two dimensional electron gas (2DEG). Sheet charge carrier density of the 2DEG in the range of $1\text{–}3 \cdot 10^{13}/\text{cm}^2$ has been determined for different Al-doping levels of the AlGaN layer by electrical measurements and can be precisely modelled [59]. This correlates to a electron sheet charge density of 1.5–4.5 $\mu\text{C/cm}^2$, a value that is similar to state-of-the-art silicon MOSFET devices. The (spontaneous and piezoelectric) polarization effects

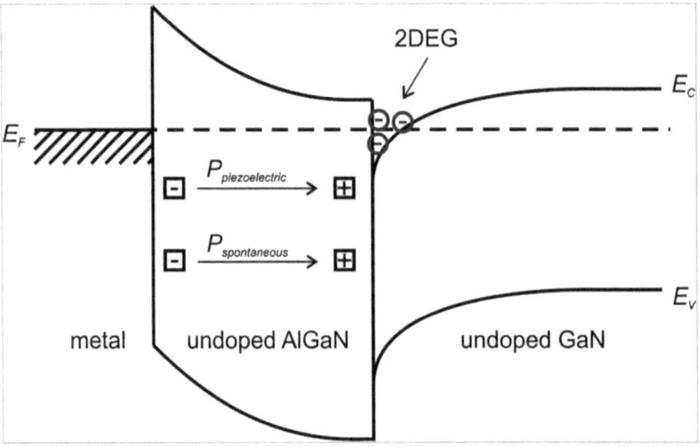

Fig. 9.18 Band diagram of AlGaN/GaN heterostructure; formation of a two dimensional electron gas due to piezoelectric and spontaneous polarization

form a 2DEG without the necessity of further doping the semiconductor layers. An electron mobility of up to 6,000 cm^2/Vs has been measured in the 2DEG [60]. Further investigations have also identified the use of AlN layers instead of the AlGaN layer as advantageous regarding mobility and sheet charge density [61]. The mobility of the 2DEG also depends on the composition of the AlGaN layer with respect to the Al/Ga ratio.

9.2.2 *Performance of High Electron Mobility Transistors*

Similar to silicon LDMOS technology, device topology of AlGaN/GaN HEMTs has evolved from the fundamental concept to specialized designed for RF amplifiers and power electronic switch-mode converters, respectively. Both traits are described in this section.

9.2.2.1 GaN Devices for Radio Frequency Operation

Early AlGaN/GaN HEMTs on sapphire substrates with 0.28 μm long gates showed an on-resistance of approximately 2.85 Ω mm with a breakdown voltage of 80 V. The transition frequency was already as high as 52 GHz [62]. The use of a SiC substrate for better heat removal and the passivation with silicon nitride enabled a further increase in stable RF operating with transition frequency of 60 GHz at a drain voltage of 7 V [63]. Advances in the growth of AlGaN/GaN epitaxial layers allowed for an increase of transition frequency of this type of device to over 120 GHz [64].

High breakdown voltage of 160 V was achieved for devices with a gate field plate resulting in a maximum oscillation frequency of 45 GHz when operated at 30 V [65]. However, the gate field plate increases gate-drain capacitance, limiting the available transistor gain. In order to overcome this limitation, dual field plate GaN HEMTs were introduced. These devices exhibit breakdown voltages of up to 250 V. A cross-section of such a device is shown in Fig. 9.19.

Further improvements to fabrication technology, the implementation of smaller epitaxial layer thicknesses and a reduction in gate length has enabled the development of AlGaN/GaN HEMTs with transition frequency of 190 GHz [67] and maximum oscillation frequency of 300 GHz [68].

The device performance of GaN HEMTs towards radio-frequency circuits is summarized in Table 9.3.

As for the 4H-SiC devices, a comparison of maximum frequency of oscillation over a wide range of breakdown voltages is presented in Fig. 9.20, yielding astonishing results.

Using GaN HEMTs, RF amplifiers with outstanding performance have been demonstrated. Operating at 4 GHz and a bias voltage of 25 V, a class AB amplifier circuit with GaN HEMT exhibited a PAE of 57.5 %, power gain of 15 dB and

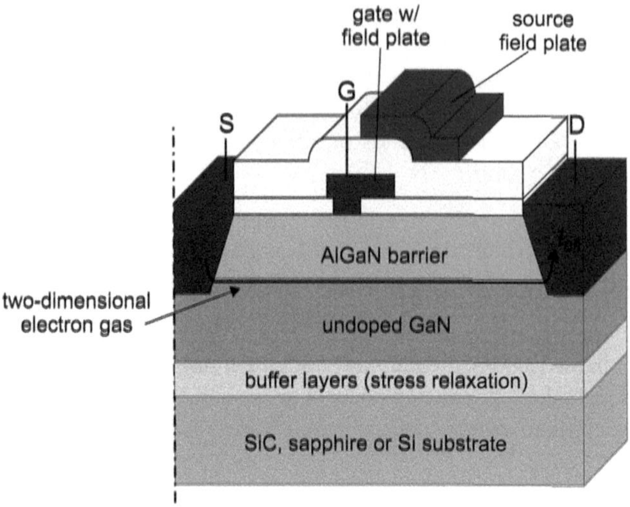

Fig. 9.19 AlGaN/GaN HEMT RF structure with gate field plate and source field plate for power electronic applications, adopted from [66]

Table 9.3 Breakdown voltage, transition and maximum oscillation frequencies for GaN HEMTs on foreign substrates

BV (V)	f_T (GHz)	f_{max} (GHz)	Device type	Substrate	Source
160	–	45	GaN HEMT	SiC	[65]
200	18	31	GaN HEMT	Si	[69]
20	70	300	GaN HEMT	SiC	[68]
40	189	251	GaN HEMT	SiC	[67]
60	75	110	GaN HEMT	SiC	[63]
60	121	162	GaN HEMT	SiC	[64]
350	4.1	6.4	GaN HEMT	Si w/Sinker	[70]
650	10	21.3	GaN HEMT	SiC	[71]
100	17.5	44	GaN HEMT	Sapphire	[62]
80	52	82	GaN HEMT	Sapphire	[62]
1,900	18.4	65	GaN HEMT	SiC	[72]
250	–	20	GaN HEMT	SiC	[66]
100	50	100	GaN HEMT	Si	[73]
95	22.7	34	GaN HEMT	Sapphire	[74]
35	74.2	–	GaN HEMT	SiC	[74]

maximum output power of 36 dBm (equals 4.12 W/mm) [74]. An output power of 41.5 dBm has been measured for a GaN HEMT in class AB mode with a power gain of 20.1 dB and a PAE of 51.4 % when biased at 40 V [66].

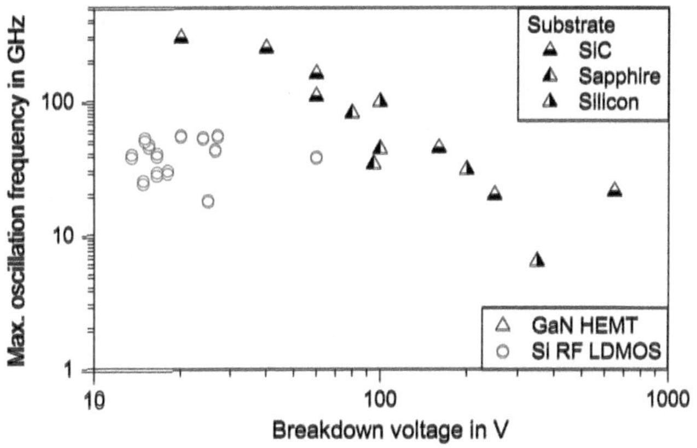

Fig. 9.20 Breakdown voltage versus maximum oscillation frequency for AlGaN/GaN HEMTs, data from Table 9.3

9.2.2.2 GaN Devices Tailored Towards Power Electronic Applications

Applicability of AlGaN/GaN HEMTs with breakdown voltages of 600 V was demonstrated by Yoshida and Ishii on sapphire substrates [75]. The device was fabricated with source and drain metallization on the AlGaN layer. An on-state resistance of 300 mΩ mm^2 for a single HEMT cell was measured. By paralleling 400 of these devices, a drain current of 15 A was obtained. However, on-state resistance increased to 7 Ω mm^2 as a large portion of the HEMT cells did not operate due to surface defects.

A 122 W Boost converter with 94.2 % efficiency including a GaN HEMT with double field plate was experimentally verified by Saito et al. [76]. The GaN HEMT showed an on-state resistance of 300 mΩ mm^2 and a breakdown voltage of 940 V. Still, the device suffers from current collapse and the device resistance increased by a factor of 4.3 when switched between on- and off-state at 350 V.

Breaking the 1 kV limit, transistors with breakdown voltages of 1,100 and 1,900 V at on-state resistances of 1.5 and 1.2 Ω mm^2 were developed [72, 77]. Implementing a surface passivation, AlGaN/GaN HEMTs on sapphire with a breakdown voltage of 1.6 kV and 340 mΩ mm^2 were also demonstrated [78].

The highest breakdown voltage for an AlGaN/GaN HEMT has been reported by Uemoto et al. [79] by using via-holes through the sapphire substrate and a 1 μm thick AlN passivation. These devices with 125 μm gate-drain distance were capable of blocking 8,300 V while offering an on-state resistance of 18.6 Ω mm^2.

Facilitating a higher thermal bulk conductivity for devices driving high current densities, GaN HEMTs grown on silicon carbide achieved an on-state resistance of 75 mΩ mm^2 at a breakdown voltage of 500 V. Switching time of these devices at a current density of 1.2 kA/cm^2 was shown to be below 5 ns [80]. Similarly, a breakdown voltage of 1,050 V with an on-state resistance of 340 mΩ mm^2 were demonstrated [81].

Xing et al. introduced multiple field plates to improve breakdown voltage of their AlGaN/GaN HEMTs. Their devices with two stacked field plates yielded a breakdown voltage of 880 V and an on-state resistance of 270 mΩ mm^2 [82]. Bahat-Treidel et al. combined a multiple graded field plate design with an AlGaN back barrier between the substrate and the GaN layer [71]. Employing a single gate field plate, they measured breakdown voltages of 550 V and an on-state resistance of 70 mΩ mm^2.

Towards the reduction of fabrication costs and the integration of GaN HEMTs in silicon technology, AlGaN/GaN HEMT fabricated on silicon with backside grounding structure (similar to the source sinker in silicon LDMOS transistors) exhibit breakdown voltages of 350 V and on-state resistance of 190 mΩ mm^2. Current densities up to 2 kA/cm^2 could be turned on and off within 98 and 96 ps, respectively. A transition frequency of 4.1 GHz was obtained [70]. Another study reports AlGaN/GaN HEMT on silicon with breakdown voltage of 1.8 kV and on-state resistance of 700 mΩ mm^2 [83].

These and further results on power GaN HEMTs have been gathered in Table 9.4.

Comparing these results, the prospects of GaN HEMTs as lateral power transistors are clearly evident from Fig. 9.21.

Clearly, channel resistance is not a limitation for GaN HEMTs in the investigated voltage range due to the high electron mobility and the charge density in the 2DEG generated by the polarization effects. The channel resistance of a GaN

Table 9.4 Breakdown voltage, on-state resistance and gate charge measured from GaN HEMTs for power electronic applications reported in literature

BV (V)	$R_{DS,on}$ (Ω mm^2)	Q_{gd} (nC/mm^2)	Device type	Substrate	Source
500	0.075	–	GaN HEMT	4H-SiC	[80]
8,300	18.6	–	GaN HEMT	Sapphire	[79]
600	0.3	–	GaN HEMT	Sapphire	[75]
350	0.19	–	GaN HEMT	Silicon	[70]
650	0.068	–	GaN HEMT	SiC	[71]
1,900	0.22	–	GaN HEMT	SiC	[72]
880	0.27	–	GaN HEMT	SiC	[82]
1,100	1.5	–	GaN HEMT	Sapphire	[77]
1,800	0.7	–	GaN HEMT	Silicon	[83]
1,050	0.34	–	GaN HEMT	SiC	[81]
435	0.4	–	Normally-off GaN HEMT	Sapphire	[84]
940	0.36	1.5	GaN HEMT	Sapphire	[85]
600	0.23	1.4	GaN HEMT	Sapphire	[85]
1,600	0.34	–	GaN HEMT	Sapphire	[78]

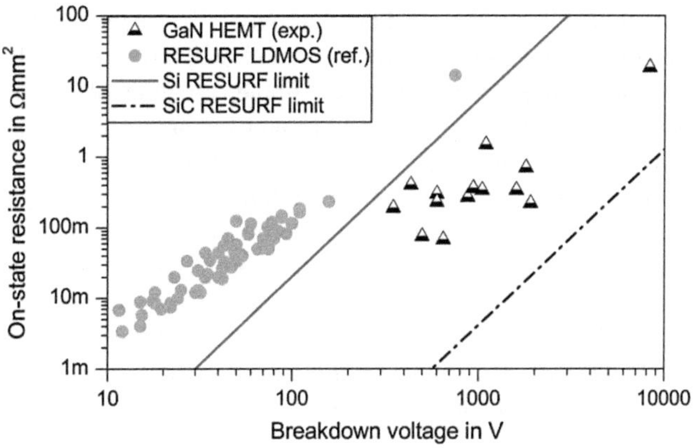

Fig. 9.21 Breakdown voltage versus on-state resistance of GaN HEMTs compared to state-of-the-art silicon LDMOS transistors, data from Table 9.4

HEMT with 1 μm channel length will be below 3 Ω mm^2 according to (7.5) and using typical values for GaN material properties. While the GaN HEMTs exceed the silicon RESURF limit, there is still room for improvement—especially towards higher breakdown voltages and the minimization of current collapse. Moreover, Fig. 9.21 does not account for the reduced polarization charge when normally-off operation is desired.

9.2.3 Technological Limitations of GaN Devices

Undesired operation of GaN-based semiconductor devices occurs due to physical effects that are related to technological limitations.

Normally-On Operation and Gate Leakage The formation of the two-dimensional electron gas in AlGaN/GaN HEMTs by polarization effects results in a conductive channel between source and drain at 0 V gate bias. This normally-on operation is not desired in power electronic applications as it can lead to a short-circuit situation in energy conversion circuit when the gate driver is disconnected during operation. Therefore, effort to achieve normally of operation has been put forward. This includes, for instance, the introduction of recessed-gate structures [84] and the implementation of gate dielectrics [86]. Using an insulated gate structure also reduced the gate leakage current of the HEMT similar to the transition from a JFET to a MOSFET structure fabricated on silicon or silicon carbide. Subsequently, the losses associated with this leakage current are reduced and higher input impedance can be achieved.

Thermal Resistance As GaN single crystals are not commercially available, substrates like sapphire and silicon carbide have to be employed. For power electronic applications, a significant self-heating was observed for devices on sapphire substrate due its low thermal conductivity [87]. As long as free-stand GaN substrates are not available, GaN power transistors will have to default to silicon or silicon carbide substrates necessitating the implementation of isolating or semi-insulating buffer layers. Or, novel assembly and mounting techniques like wafer grinding or front-side cooling need to be further developed.

Dynamic On-Resistance and Current Collapse Reduction of output current of AlGaN/GaN HEMTs has been observed at microwave frequencies [88]. This effect diminishes the linearity of RF amplifiers based on these devices. Further investigations indicate that piezo-related surface states are responsible for this current dispersion. A suitable surface passivation shifts this effect towards higher operating frequencies [89]. Moreover, a source field plate has been successfully employed to reduce the impact of current collapse by reducing the electric field at the edge of the gate electrode [85].

9.2.4 GaN-Based MMICs and Class-S Power Amplifiers

MMICs for RF power amplifiers employing GaN HEMTs are about to be introduced to the market. A review of GaN-on-SiC based MMICs from the point of view of a semiconductor company is given in [90]. GaN HEMTs have enabled the design and fabrication of switch-mode based amplifiers with transition frequencies and breakdown voltages significantly higher than in silicon LDMOS based solutions.

GaN HEMTs have enabled the implementation of switch-mode based power amplifier concepts achieving high energy efficiencies. A 10 W class S amplifier operating from 0.55 to 1.1 GHz with an average drain efficiency of 74 % was demonstrated [91].

For satellite communication and radar applications, three-stage GaN MMICs have been realized. Power amplifiers based on this design employ transistors with transition frequency of 90 GHz at 20 V bias [92]. Current versions of these amplifiers provide an output power of up to 840 mW at 88 GHz and a power added efficiency of 15 % for 782 mW output power [93].

9.2.5 Integration Consideration for GaN in Smart-Power ICs

Clearly, the incorporation of lateral GaN HEMTs in smart-power IC technology would constitute to the evolution of a high-power, high-voltage IC technology. For this purpose, the feasibility of digital circuit technology using GaN HEMTs has been shown even at operating temperature of up to 300 °C [94]. Unfortunately, free-standing GaN material is not available in wafer diameters suitable for commercial circuit fabrication.

There is extensive work being carried out towards the co-integration of InGaAs-based n-channel MOSFETs into standard CMOS technology for future CMOS generations [95]. In order to reduce fabrication cost for GaN material to provide large wafer diameters for fabrication of GaN devices, GaN-on-Si development has gained momentum over the last decade. The growth of GaN films on silicon substrates has already been demonstrated in the 1990s and different intermediate layers, e.g. 3C-SiC [96] and AlN [97], have been investigated to cope with the large lattice mismatch between GaN and silicon. Recently, work on the fabrication of AlGaN/GaN HEMTs on silicon has been presented that is based on a CMOS compatible process flow [98]. However, up-to-date no high-quality 200 mm GaN-on-Si wafer material is available which could be included into the fabrication sequence of a silicon-based MOS process for commercialization.

9.3 Lateral Power Transistors on Wide Bandgap Semiconductors Fabricated in Integrated Circuits

In summary, both silicon carbide and gallium nitride achieve outstanding device performance for power electronic and radio-frequency applications. However, the ability to integrate these devices into silicon technology does currently not exist. For silicon carbide, vertical power transistors are available for fabrication of efficient power converters. For gallium nitride, lateral GaN HEMTs are employed as stand-alone devices in RF amplifiers.

Simple CMOS logic has been investigated and demonstrated in 4H-SiC which could be an option for high temperature and/or radiation hard smart-power IC technology in particularly demanding niche markets. Similarly, JFET-based logic and power devices could alleviate smart-power IC in SiC due to the lack of a critical gate oxide.

For GaN-based smart-power ICs, GaN-on-Si technology on 150 and 200 mm wafers could quickly enable CMOS compatible processing in existing foundries to roll out new products that could provide RF functionality in parallel. However, this highly versatile solution is still a vision.

References

1. R. Davis, GaAs FETs—physics, design, and models, in *Handbook of RF and Microwave Power Amplifiers* (Cambridge University Press, Cambridge, 2012), pp. 42–102
2. G.I. Gudjonsson, F. Allerstam, E.Ö. Sveinbjörnsson, H. Hjelmgren, P.-A. Nilson, K. Andersson, H. Zirath, T. Rödle, R. Jos, Design and fabrication of 4H-SiC RF MOSFETs. IEEE Trans. Electron Devices **54**(12), 3138–3145 (2007)
3. D.P. Myer, Practical HF/VHF/UHF RF power amplifier realization, in *Handbook of RF and Microwave Power Amplifiers*, (Cambridge University Press, Cambridge, 2012), pp. 232–283
4. S.-H. Ryu, S. Dhar, S. Haney, A. Agarwal, A. Lelis, B. Geil, C. Scozzie, Critical issues for MOS based power devices in 4H-SiC. Mater. Sci. Forum. **743–748**, 743–748 (2009)
5. W. Fulop, Calculation of avalanche breakdown of silicon p-n junctions. Solid-State Electron. **10**(1), 39–43 (1967)

6. B.J. Baliga, *Fundamentals of Power Semiconductor Devices* (Springer, New York, 2008)
7. I.A. Khan, J.A. Cooper, Measurement of high-field electron transport in silicon carbide. IEEE Trans. Electron Devices **47**(2), 269–273 (2000)
8. M. Farahmand, C. Garetto, E. Bellotti, K. Brennan, M. Goano, E. Ghillino, G. Ghione, J.D. Albrecht, P.P. Ruden, Monte Carlo simulation of electron transport in the III-nitride wurtzite phase materials system: binaries and ternaries. IEEE Trans. Electron Devices **48**(3), 535–542 (2001)
9. F. Schwierz, An electron mobility model for wurtzite GaN. Solid-State Electron. **49**(6), 889–895 (2005)
10. C. Canali, G. Majni, R. Minder, G. Ottaviani, Electron and hole drift velocity measurements in silicon and their empirical relation to electric field and temperature. IEEE Trans. Electron Devices **22**(11), 1045–1047 (1975)
11. W.V. Muench, E. Pettenpaul, Saturated electron drift velocity in 6H silicon carbide. J. Appl. Phys. **48**, 4823–4825 (1977)
12. S. Takagi, A. Toriumi, M. Iwase, H. Tango, On the universality of inversion layer mobility in Si MOSFET's: part I-effects of substrate impurity concentration. IEEE Trans. Electron Devices **41**(12), 2357–2362 (1994)
13. G. Gudjonsson, H.Ö. Olafsson, F. Allerstam, P.-A. Nilson, E.Ö. Sveinbjörnsson, H. Zirath, T. Rödle, R. Jos, High field-effect mobility in n-channel Si face 4H-SiC MOSFETs with gate oxide grown on aluminium ion-implanted material. IEEE Electron Device Lett. **26**(2), 96–98 (2005)
14. M.A. Green, Intrinsic concentration, effective densities of states, and effective mass in silicon. J. Appl. Phys. **67**(6), 2944–2954 (1990)
15. J.B. Casady, R.W. Johnson, Status of silicon carbide (SiC) as a wide-bandgap semiconductor for high-temperature applications: a review. Solid-State Electron. **39**(10), 1409–1422 (1996)
16. S.K. Laha, M. Mukherjee, Physics based modeling of intrinsic material parameters of III–V nitride semiconductors: elevated temperature effects. Int. J. Recent Technol. Eng. **2**(6), 48–50 (2014)
17. F.P. McCluskey, R. Grzybowski, T. Podlesak, *High Temperature Electronics* (CRC Press, Boca Raton, 1997)
18. S.M. Sze, K.K. Ng, *Physics of Semiconductor Devices* (John Wiley, Hoboken, 2007)
19. J.W. Palmour, H.S. Kong, D.G. Waltz, J.A. Edmond, C.H. Carter Jr., 6H silicon carbide transistors for high temperature operation, in *International High Temperature Electronics Conference*, Ottawa, Canada, 1991
20. G.L. Harris, Thermal conductivity of SiC, in *Properties of Silicon Carbide* (INSPEC, The Institution of Electric Engineers, London, 1995), pp. 5–6
21. C.J. Glassbrenner, G.A. Slack, Thermal conductivity of silicon and germanium from 3 °K to the melting point. Phys. Rev. **134**(4A), A1058–A1069 (1964)
22. X. Wang, Y. Ezzahri, J. Christofferson, A. Shakouri, Bias-dependent MOS transistor thermal resistance and non-uniform self-heating temperature. J. Phys. D Appl. Phys. **42**, 075101 (2009)
23. R.J. Trew, *Wide band gap transistors—SiC and GaN—physics, design and models*, ed. by John L. B. Walker. Handbook of RF and Microwave Power Amplifiers (Cambridge University Press, Cambridge, 2012), pp. 103–158
24. C.E. Weitzel, Comparison of SiC, GaAs, and Si R F MESFET power densities. IEEE Electron Device Lett. **16**(10), 451–453 (1995)
25. S.T. Allen, J.W. Palmour, V.F. Tsvetkov, S.J. Macko, C.H. Carter Jr., C.E. Weitzel, K.E. Moore, K.J. Nordquist, L.L. Pond III, 4H-SiC MESFET's on high resistivity substrates with 30 GHz fmax, in *Device Research Conference*, Charlottesville, USA, 1995
26. S. Sriram, G. Augustine, A.A. Burk Jr, R.C. Glass, H.M. Hobgood, P.A. Orphanos, L.B. Rowland, T.J. Smith, C.D. Brandt, M.C. Driver, R.H. Hopkins, 4H-SiC MESFET's with 42 GHz fmax. IEEE Electron Device Lett. **17**(7), 369–371 (1996)
27. R.J. Trew, SiC and GaN transistors—is there one winner for microwave power applications? Proc. IEEE **90**(6), 1032–1047 (2002)

28. O. Noblanc, E. Chartier, C. Arnodo, C. Brylinski, Microwave power MESFET on 4H-SiC. Diam. Relat. Mater. **6**, 1508–1511 (1997)
29. S. Sriram, H. Hagleitner, D. Namishia, T. Alcorn, T. Smith, B. Pulz, High-gain SiC MESFETs using source-connected field plates. IEEE Electron Device Lett. **30**(9), 952–953 (2009)
30. F. Temcamani, P. Pouvil, O. Noblanc, C. Brylinski, P. Bannelier, B. Darges, J.P. Prigent, Silicon carbide MESFETs performances and application in broadcast power amplifiers, in *IEEE MTT-S International Microwave Symposium*, Phoenix, USA, 2001
31. D. Alok, E. Arnold, R. Egloff, J. Barone, J. Murphy, R. Conrad, J. Burke, 4H-SiC RF power MOSFETs. IEEE Electron Devices Lett. **22**(12), 577–578 (2001)
32. C. Strenger, V. Häublein, T. Erlbacher, A.J. Bauer, H. Ryssel, A.M. Beltran, S. Schamm-Chardon, V. Mortet, E. Bedel-Pereira, M. Lefebvre, F. Cristiano, Comparative study of electrical and microstructural properties of 4H-SiC MOSFETs. Mater. Sci. Forum **717–720**, 437–440 (2012)
33. G. Gudjonsson, F. Allerstam, H.Ö. Olafsson, P.A. Nilsson, H. Hjelmgren, K. Andersson, E.Ö. Sveinbjörnsson, H. Zirath, T. Rödle, R. Jos, High-power-density 4H-SiC RF MOSFETs. IEEE Electron Device Lett. **27**(6), 469–471 (2006)
34. G.I. Gudjonsson, F. Allerstam, E.Ö. Sveinbjörnsson, H. Hjelmgren, P.-A. Nilsson, K. Andersson, H. Zirath, T. Rödle, R. Jos, Design and fabrication of 4H-SiC RF MOSFETs. IEEE Trans. Electron Devices **54**(12), 3138–3145 (2007)
35. J. Spitz, M.R. Melloch, J.A. Cooper Jr, M.A. Capano, 2.6 kV 4H-SiC lateral DMOSFET's. IEEE Electron Device Lett. **19**(4), 100–102 (1998)
36. K. Chatty, S. Banerjee, T.P. Chow, R.J. Gutmann, M. Hoshi, High-voltage lateral RESURF MOSFETs on 4H-SiC, in *Device Research Conference*, Santa Barbara, USA, 1999
37. S. Bannerjee, K. Chatty, T.P. Chow, R.J. Gutmann, Improved high-voltage lateral RESURF MOSFETs in 4H-SiC. IEEE Electron Device Lett. **22**(5), 209–211 (2001)
38. V. Häublein, G. Temmel, H. Mitlehner, G. Rattmann, C. Strenger, A. Hürner, A.J. Bauer, H. Ryssel, L. Frey, Comparative study of n-LIGBT and n-LDMOS structures on 4H-SiC. Mater. Sci. Forum **740–742**, 887–890 (2013)
39. M.B.J. Wijesundra, R. Azevedo, *Silicon Carbide Microsystems for Harsh Environments* (Springer, New York, 2011)
40. B. Kallinger, B. Thomas, S. Polster, P. Berwian, J. Friedrich, Dislocation conversion and propagation during homoepitaxial growth of 4H-SiC. Mater. Sci. Forum **645–648**, 299–302 (2010)
41. T. Troffer, M. Schadt, T. Frank, H. Itoh, G. Pensl, J. Heindl, H.P. Strunk, M. Maier, Doping of SiC by implantation of boron and aluminum. Phys. Status Solidi A **162**, 277–298 (1997)
42. N.S. Saks, A.V. Suvorov, D.C. Capell, High temperature high-dose implantation of aluminum in 4H-SiC. Appl. Phys. Lett. **84**(25), 5195–5197 (2004)
43. N.S. Saks, S.S. Mani, A.K. Agarwal, Interface trap profile near the band edges at the 4H-SiC/SiO2 interface. Appl. Phys. Lett. **76**, 2250–2252 (2000)
44. R.R. Razouk, B.E. Deal, Dependence of interface state density on silicon thermal oxidation process variables. J. Electrochem. Soc. **126**(9), 1573–1581 (1979)
45. K. Kita, R.H. Kikuchi, H. Hirai, Understanding of growth kinetics of thermal oxides on 4H-SiC (0001) for control of MOS characteristics. ECS Trans. **61**(2), 135–142 (2014)
46. J.A. Sprague, A. Addamiano, "Bufferlayer" technique for the growth of single crystal SiC on Si. Appl. Phys. Lett. **44**, 525–527 (1984)
47. W. Xie, J.A. Cooper, M.R. Melloch, Monolithic NMOS digital integrated circuits in 6H-SiC. IEEE Electron Device Lett. **15**(11), 455–457 (1994)
48. J.-S. Chen, S.-H. Ryu, K.T. Kornegay, A silicon carbide CMOS intelligent gate driver circuit. IEEE J. Solid-State Circuits **34**(2), 192–204 (1999)
49. P.G. Neudeck, D.J. Spry, L.-Y. Chen, G.M. Beheim, R.S. Okojie, C.W. Chang, R.D. Meredith, T.L. Ferrier, L.J. Evans, M.J. Krasowski, N.F. Prokop, Stable electrical operation of 6H-SiC JFETs and ICs for thousands of hours at 500 °C. IEEE Electron Device Lett. **29**(5), 456–459 (2008)

50. I. Sankin, V. Bondarenko, R. Kelley, J.B. Casady, SiC smart power JFET technology for high-temperature applications. Mater. Sci. Forum **527–529**, 1207–1210 (2006)
51. Y. Zhang, K. Sheng, M. Su, J.H. Zhao, P. Alexandrov, X. Li, L. Fursin, M. Weiner, Development of 4H-SiC LJFET-based power IC. IEEE Trans. Electron Devices **55**(8), 1934–1945 (2008)
52. A. Devie, D. Tournier, P. Godignon, M. Vellvehi, J. Montserrat, X. Jorda, Fabrication and testing of 4H-SiC MESFETs for analog functions circuits. Mater. Sci. Forum **645–648**, 1159–1162 (2010)
53. M. Le-Huu, M. Grieb, F.F. Schrey, H. Schmitt, V. Häublein, A.J. Bauer, H. Ryssel, L. Frey, 4H-SiC n-MOSFET logic circuits for high temperature operation. Mater. Sci. Forum **679–680**, 734–737 (2011)
54. W. Daves, A. Krauss, M. Le-Huu, S. Kronmüller, V. Häublein, A.J. Bauer, L. Frey, Comparative study on metallization/passivation technologies for high temperature sensor applications. Mater. Sci. Forum **679–680**, 449–452 (2011)
55. Z. Stum, V. Tilak, P. Losee, E. Andarawis, C. Chen, 300 °C Silicon carbide integrated circuits. Mater. Sci. Forum **679–680**, 730–733 (2011)
56. A. Patil, N. Rao, V. Tilak, 300 °C Capable digital integrated circuits in SiC technology. Mater. Sci. Forum **717–720**, 1261–1264 (2012)
57. B.A. Hull, S.-H. Ryu, H. Fatima, J. Richmond, J.W. Palmour, J. Scofield, Development of a 4H-SiC CMOS inverter, in *MRS Proceedings*, pp. B13-02, 2006
58. D.T. Clark, E.P. Ramsay, A.E. Murphy, D.A. Smith, R.F. Thompson, R.A.R. Young, J.D. Cormack, C. Zhu, S. Finney, J. Fletcher, High temperature silicon carbide CMOS integrated circuits. Mater. Sci. Forum **679–680**, 726–729 (2011)
59. O. Ambacher, B. Foutz, J. Smart, J.R. Shealy, N.G. Weimann, K. Chu, M. Murphy, A.J. Sierakowski, W.J. Schaff, L.F. Eastman, Two dimensional electron gases induced by spontaneous and piezoelectric polarization in undoped and doped AlGaN/GaN heterostructures. J. Appl. Phys. **87**(1), 334–344 (2000)
60. Z. Dziuba, J. Antoszewski, J.M. Dell, L. Faraone, P. Kozodoy, S. Keller, B. Keller, S.P. DenBaars, U.K. Mishra, Magnetic field dependent Hall data analysis of electron transport in modulation-doped AlGaN/GaN heterostructures. J. Appl. Phys. **82**, 2996–3002 (1997)
61. C. Yu, D. Jena, High-mobility window for two-dimensional electron gases at ultrathin AlN/GaN heterojunctions. Appl. Phys. Lett. **90**(18), 182112 (2007)
62. U.K. Mishra, Y.-F. Wu, B.P. Keller, S. Keller, S.P. Denbaars, GaN microwave electronics. IEEE Trans. Microw. Theory Tech. **46**(6), 756–761 (1998)
63. R. Quay, R. Kiefer, F. van Raay, H. Massler, S. Ramberger, S. Müller, M. Dammann, M. Milulla, M. Schlechtweg, G. Weimann, AlGaN/GaN HEMTs on SiC operating at 40 GHz, in *International Electron Devices Meeting*, San Francisco, USA, 2002
64. V. Kumar, W. Lu, R. Schwindt, A. Kuliev, G. Simin, J. Yang, M.A. Khan, I. Adesida, AlGaN/GaN HEMTs on SiC with ft of over 120 GHz. IEEE Electron Device Lett. **23**(8), 455–457 (2002)
65. Y. Ando, Y. Okamoto, H. Miyamoto, T. Nakayama, T. Inoue, M. Kuzuhara, 10-W/mm AlGaN-GaN HFET with a field modulating plate. IEEE Electron Device Lett. **24**(5), 289–291 (2003)
66. Y. Ando, A. Wakejima, Y. Okamoto, T. Nakayama, K. Ota, K. Yamanoguchi, Y. Murase, K. Kasahara, K. Matsunaga, T. Inoue, H. Miyamoto, Novel AlGaN/GaN dual-field-plate FET with high gain, increased linearity and stability, in *International Electron Devices Meeting*, Washington, USA, 2005
67. M. Higashiwaki, T. Mimura, T. Matsui, AlGaN/GaN heterostructure field-effect transistor on 4H-SiC substrates with current-gain cutoff frequency of 190 GHz. Appl. Phys. Express **1**, 021103 (2008)
68. J.W. Chung, W.E. Hoke, E.M. Chumbes, T. Palacios, AlGaN/GaN HEMT with 300-GHz fmax. IEEE Electron Device Lett. **31**(3), 195–197 (2010)

69. J.W. Johnson, E.L. Piner, A. Vescan, R. Therrien, P. Rajagopal, J.C. Roberts, J.D. Brown, S. Singhal, K.J. Linthicum, 12 W/mm AlGaN-GaN HFETs on silicon substrates. IEEE Electron Device Lett. **25**(7), 459–461 (2004)
70. M. Hikita, M. Yanagihara, K. Nakazawa, H. Ueno, Y. Hirose, T. Ueda, Y. Temoto, T. Tanaka, D. Ueda, T. Egawa, AlGaN/GaN power HFET on silicon substrate with source-via grounding (SVG) structure. IEEE Trans. Electron Devices **52**(9), 1963–1968 (2005)
71. E. Bahat-Treidel, O. Hilt, F. Brunner, V. Sidorov, J. Würfl, G. Tränkle, AlGaN/GaN/AlGaN DH-HEMTs breakdown voltage enhancement using multiple grating field plates (MGFPs). IEEE Trans. Electron Devices **57**(6), 1208–1216 (2010)
72. Y. Dora, A. Chakraborty, L. McCarthy, S. Keller, S.P. DenBaars, U.K. Mishra, High breakdown voltage achieved on AlGaN/GaN HEMTs with integrated slant field plates. IEEE Electron Device Lett. **27**(9), 713–715 (2006)
73. B.M. Green, K.K. Chu, E.M. Chumbes, J.A. Smart, J.R. Shealy, L.F. Eastman, The effect of surface passivation on the microwave characteristics of undoped AlGaN/GaN HEMT's. IEEE Electron Device Lett. **21**(6), 268–270 (2000)
74. L.F. Eastman, V. Tilak, J. Smart, B.M. Green, E.M. Chumbes, R. Dimitrov, H. Kim, O.S. Ambacher, N. Weimann, T. Prunty, M. Murphy, W.J. Schaff, J.R. Shealy, Undoped AlGaN/GaN HEMTs for microwave power amplification. IEEE Trans. Electron Devices **48**(3), 479–485 (2001)
75. S. Yoshida, H. Ishii, A high-power GaN-based field-effect transistor for large-current operation. Phys. Status Solidi A **188**(1), 243–246 (2001)
76. W. Saito, T. Nitta, Y. Kakiuchi, Y. Saito, K. Tsuda, I. Omura, M. Yamaguchi, A 120-W boost converter operation using a high-voltage GaN-HEMT. IEEE Electron Device Lett. **29**(1), 8–10 (2008)
77. S. Yagi, M. Shimizu, H. Okumura, H. Ohashi, Y. Yano, N. Akutsu, High breakdown voltage AlGaN/GaN metal–insulator–semiconductor high-electron-mobility transistor with TiO_2/SiN gate insulator. Jpn. J. Appl. Phys. **46**(4B), 2309–2311 (2007)
78. N. Tipirneni, A. Koudymov, V. Adivarahan, J. Yang, G. Simin, M.A. Khan, The 1.6-kV AlGaN/GaN HFETs. IEEE Electron Device Lett. **27**(9), 716–718 (2006)
79. Y. Uemoto, D. Shibata, M. Yanagihara, H. Ishida, H. Matsuo, S. Nagai, N. Batta, M. Li, T. Ueda, T. Tanaka, D. Ueda, 8300 V blocking voltage AlGaN/GaN power HFET with thick poly-AlN passivation, in *International Electron Devices Meeting*, Washington, USA, 2007
80. G. Simin, X. Hu, N. Ilinskaya, A. Kumar, A. Koudymov, J. Zhang, M.A. Khan, R. Gaska, M. S. Shur, 7.5 kW/mm^2 current switch using AlGaN/GaN metal-oxide-semiconductor heterostructure field effect transistors on SiC substrates. Electron. Lett. **36**(24), 2043–2044 (2000)
81. N.-Q. Zhang, B. Moran, S.P. DenBaars, U.K. Mishra, X.W. Wang, T.P. Ma, Kilovolt AlGaN/GaN HEMTs as switching devices. Phys. Status Solidi A **188**(1), 213–217 (2001)
82. H. Xing, Y. Dora, A. Chini, S. Heikman, S. Keller, U.K. Mishra, High breakdown voltage AlGaN-GaN HEMTs achieved by multiple field plates. IEEE Electron Device Lett. **25**(4), 161–163 (2004)
83. N. Ikeda, S. Kaya, J. Li, Y. Sato, S. Kato, S. Yoshida, High power AlGaN/GaN HFET with a high breakdown voltage of over 1.8 kV on 4 inch Si substrates and the suppression of current collapse, in *International Symposium on Power Semiconductor Devices & ICs*, Orlando, USA, 2008
84. W. Saito, Y. Takada, M. Kuraguchi, K. Tsuda, I. Omura, Recessed-gate structure approach toward normally off high-voltage AlGaN/GaN HEMT for power electronics applications. IEEE Trans. Electron Devices **53**(2), 356–362 (2006)
85. W. Saito, T. Nitta, Y. Kakiuchi, Y. Saito, K. Tsuda, I. Omura, M. Yamaguchi, Suppression of dynamic on-resistance increase and gate charge measurements in high-voltage GaN-HEMTs with optimized field-plate structure. IEEE Trans. Electron Devices **54**(8), 1825–1830 (2007)
86. S. Taking, D. MacFarlane, E. Wasige, AlN/GaN-based MOS-HEMT technology: processing and device results. Act. Passive Electron. Compon. **2011**, 821305 (2011)

87. R. Gaska, A. Osinsky, J.W. Yang, M.S. Shur, Self-heating in high-power AlGaN-GaN HFET's. IEEE Electron Device Lett. **19**(3), 89–91 (1998)
88. C. Nguyen, N.X. Nguyen, D.E. Grider, Drain current compression in GaN MODFETs under large-signal modulation at microwave frequencies. Electron. Lett. **35**(16), 1380–1382 (1999)
89. I. Daumiller, D. Theron, C. Gaquiere, A. Vescan, R. Dietrich, A. Wieszt, H. Leier, R. Vetury, U.K. Mishra, I.P. Smorchkova, S. Keller, N.X. Nguyen, C. Nguyen, E. Kohn, Current instabilities in GaN-based devices. IEEE Electron Device Lett. **22**(2), 62–64 (2001)
90. R.S. Pengelly, S.M. Wood, J.W. Milligan, S.T. Sheppard, W.L. Pribble, A review of GaN on SiC high electron-mobility power transistors and MMICs. IEEE Trans. Microw. Theory Tech. **60**(6), 1764–1783 (2012)
91. V. Carrubba, J. Lees, J. Benedikt, P.J. Tasker, S.C. Cripps, A novel highly efficient broadband continuous class-F RFPA delivering 74 % average efficiency for an octave bandwidth, in *IEEE MTT-S International Microwave Symposium*, Baltimore, USA, 2011
92. M. Micovic, A. Kurdoghlian, P. Hashimoto, M. Hu, M. Antcliffe, P.J. Willadsen, W.S. Wong, R. Bowen, I. Milosavljevic, A. Schmitz, M. Wetzel, D.H. Chow, GaN HFET for W-band power applications, in *International Electron Devices Meeting*, San Francisco, USA, 2006
93. M. Micovic, A. Kurdoghlian, K. Shinohara, S. Burnham, I. Milosavljevic, M. Hu, A. Corrion, A. Fung, R. Lin, L. Samoska, P. Kangaslahti, B. Lambrigtsen, P. Goldsmith, W.S. Wong, A. Schmitz, P. Hashimoto, P.J. Willadsen, D.H. Chow, W-Band GaN MMIC with 842 mW output power at 88 GHz, in *IEEE MTT-S International Microwave Symposium*, Anaheim, USA, 2010
94. T. Hussain, M. Micovic, T. Tsen, M. Delaney, D. Chow, A. Schmitz, P. Hashimoto, D. Wong, J.S. Moon, M. Hu, J. Duvall, D. McLaughlin, GaN HFET digital circuit technology for harsh environments. Electron. Lett. **39**(24), 1708–1709 (2003)
95. L. Czornomaz, N. Daix, K. Cheng, D. Caimi, C. Rossel, K. Lister, M. Sousa, J. Fompeyrine, Co-integration of InGaAs n- and SiGe p-MOSFETs into digital CMOS circuits using hybrid dual-channel ETXOI substrates, in *International Electron Devices Meeting*, San Francisco, USA, 2012
96. T. Takeuchi, H. Amano, K. Hiramatsu, N. Sawaki, I. Akasaki, Growth of single crystalline GaN film on Si substrate using 3C-SiC as an intermediate layer. J. Cryst. Growth **115**, 634–638 (1991)
97. A. Watanabe, T. Takeuchi, K. Hirosawa, H. Amano, K. Hiramatsu, I. Akasaki, The growth of single crystalline GaN on a Si substrate using AlN as an intermediate layer. J. Cryst. Growth **128**(1–4), 391–396 (1993)
98. M. Van Hove, S. Boulay, S.R. Bahl, S. Stoffens, X. Kang, D. Wellekens, K. Geens, A. Delabie, S. Decoutere, CMOS process-compatible high-power low-leakage AlGaN/GaN MISHEMT on silicon. IEEE Electron Device Lett. **33**(5), 667–669 (2012)

Chapter 10
Summary of Integration Concepts for LDMOS Transistors

The final chapter of this textbook summarizes the integration concepts presented in Chap. 5 through Chap. 9. Comparison of integration densities and figures-of-merit represent a first methodology to evaluate these concepts with respect to applications requiring high efficiency or performance. Another methodology for comparison of device technologies is the comparison of process complexity and fabrication effort which is a key indicator for cost-sensitive applications. Using these comparisons, feasibility of emerging device technologies can be investigated with respect to different applications. The application-specific suitability is exemplified with the applications and circuits presented in Chaps. 2 and 3.

10.1 Integration Density and Figures-of-Merit

The improvement of lateral power transistors in lateral power devices is of utmost importance for further advances in integrated power electronic and radio frequency applications.

10.1.1 Devices for Power Electronic Systems

In power electronics, a reduction of on-state resistance is desired. It allows for a reduction in both static power losses and chip size. The More-than-Moore approaches discussed in Chap. 5 through Chap. 9 are compared regarding the on-state resistance and the breakdown voltage in Fig. 10.1.

The superjunction LDMOS transistors published in literature achieve similar values of on-state resistance as the RESURF LDMOS transistors that are considered state-of-the-art. Even though drift region resistance of superjunction patterns can be lower than the RESURF limit, the high depth-to-width ratios associated with this are difficult to fabricate, e.g. requiring high energy implantation. The superjunction devices cannot excel over the RESURF LDMOS transistors at high breakdown voltages either. Basically, the RESURF design is a lateral superjunction pattern and

T. Erlbacher, *Lateral Power Transistors in Integrated Circuits*,
Power Systems, DOI 10.1007/978-3-319-00500-3_10

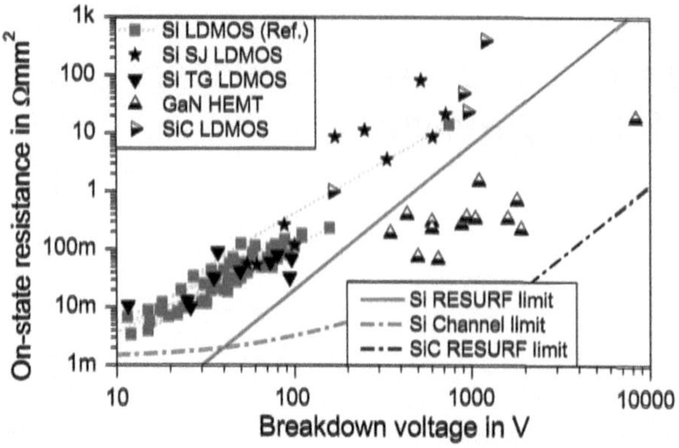

Fig. 10.1 Comparison of breakdown voltage and on-state resistance for different semiconductor materials and device topologies

both concepts rely on the depletion of the drift region(s) from the direction perpendicular to current flow.

Trench gate LDMOS transistors are an innovative concept for transistors with breakdown voltages of 100 V and below, where the contribution of the channel resistance to the total on-state resistance is dominant. Here, the trench gate integration allows for a higher density of channels than for planar technology. Moreover, one trench provides two channel regions (one on each side of the trench). Significant reduction of on-state resistance can be achieved, but the integration of trench gates will likely be limited to special cases where trench gate integration can be readily applied. Nevertheless, FinFET devices will be available when today's high-end CMOS technologies will eventually become available for smart-power IC fabrication.

SiC LDMOS transistors are still suffering from a low inversion mobility which has prevented the realization of low-loss lateral power transistors. Nevertheless, the capability of vertical SiC transistors has been proven. With further improvements to gate oxidation and an increase in inversion mobility, lateral SiC transistors can easily break the silicon RESURF limit at voltages of 650 V and above.

Finally, GaN HEMTs exhibit very promising drift region resistance as lateral power devices. Without a doubt, vertical SiC devices are challenged by GaN HEMTs for their low device resistance. However, these devices are usually normally-on limiting their acceptance among circuit designers. Moreover, integration of GaN HEMTs in silicon technology relies on the fabrication of GaN-on-silicon technology that is not yet available for wafer diameters used in foundries. Integration of GaN transistors in integrated circuits is barely investigated.

Low dynamic power losses are incurred when the device switches fast. Switching speed is—among other reasons—limited by the transfer capacitance (Miller capacitance) of the transistor. A comparison of gate charge values for different LDMOS technologies is given in Fig. 10.2.

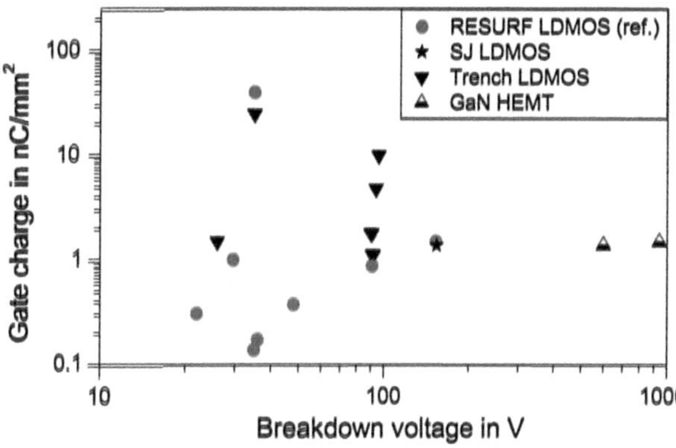

Fig. 10.2 Comparison of breakdown voltage and gate charge density for different semiconductor materials and device topologies

While data regarding gate charge is scarce in general, the low values of GaN HEMTs are apparent. A field plate is omitted in most GaN HEMTs for RF applications presented in literature. Moreover, any field plate that has to be present for proper device performance (including avoidance of current collapse) can be shorter than in silicon due to the reduced drift region length. For SiC LDMOS devices, no data was found in literature but similar considerations as for the GaN HEMTs should hold true as long as the electric field across the gate oxide can be limited by other means, e.g. implanted p^+-doping shield region.

Lastly, trench gate LDMOS transistors are typically subject to gate charge values higher than RESURF transistors. The transfer capacitance increases for a higher gate region density. Moreover, the channel length of a trench gate is usually larger in integrated circuits, because the depth of the wells typically exceeds the length of a planar gate.

10.1.2 Devices for Radio-Frequency Amplifiers

In amplification at radio frequencies, high transistor gain at high frequencies is desired. Considering the gain-bandwidth product, the maximum oscillation frequency is a first indicator of RF performance. The maximum oscillation frequencies of lateral power transistors fabricated with the concepts and materials from Chap. 5 through Chap. 9 are summarized in Fig. 10.3.

Silicon RF transistors are limited to breakdown voltages below 100 V. Maximum oscillation frequencies of 40–60 GHz can be achieved in integrated circuits using this established technology. The trench gate RF transistors are not up to par due to the higher device capacitances in trench gate transistors as previously discussed. This may change when FinFET-type devices will be fabricated for RF CMOS applications.

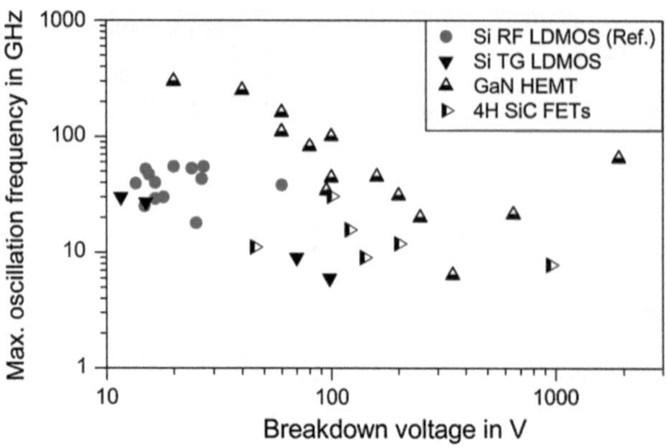

Fig. 10.3 Comparison of breakdown voltage and maximum oscillation frequencies for different semiconductor materials and device topologies

Higher gate-drain capacitance limits application of superjunction patterns in RF LDMOS transistors [1].

SiC-based field effect transistors enable device operation at RF at several hundred volts. Their maximum oscillation frequency is similar to silicon devices.

The two-dimensional electron gas in GaN HEMTs enables excellent RF performance. Maximum oscillation frequency and breakdown voltage exceed the values of silicon RF transistors by almost a factor of ten. As the switching frequency also strongly depends on the gate length, a comparison of transition frequency on gate length for the technologies discussed in Chap. 5 through Chap. 9 is presented in Fig. 10.4.

The extrapolation of the average transition frequency for GaN HEMTs towards shorter gate length clearly indicates the potential for GaN-based RF MMICs. For the other technologies, an extrapolation to shorter channel length and a comparison

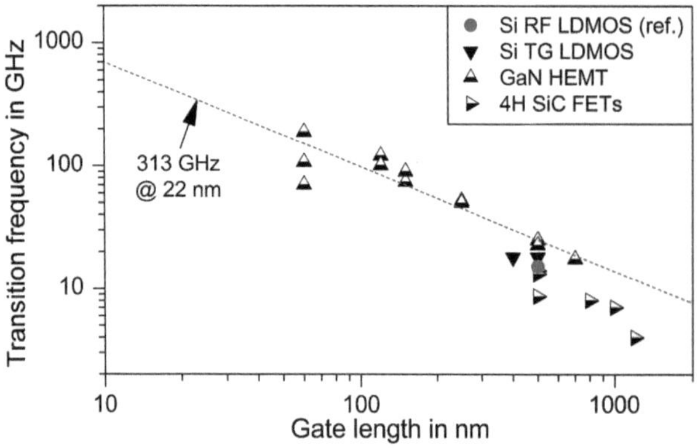

Fig. 10.4 Transition frequency dependence on gate length; extrapolation of GaN HEMT data to short gate lengths

to state-of-the-art silicon LDMOS transistors is difficult due to the small data base. Estimation may be possible in the future as devices fabricated using smaller technology nodes become available.

10.2 Process Complexity and Cost

The incorporation of advanced device topologies also requires additional fabrication effort. This has to be justified by improved performance, i.e. enabling a new system approach or a reduction of chip area for a given application. In order to judge the effort for the technologies previously discussed compared to conventional single- and double-RESURF LDMOS transistors, the additional mask count and process modules for these technologies are shown in Table 10.1.

Some of these figures depend on the reusability of existing process modules and photo mask layers of the CMOS fabrication processes. It is feasible to explore the multi-RESURF design and superjunction patterns if no additional photo mask layers are required and implantation may be reused. However, this limits the benefits that can be achieved compared to a more flexible design with additional mask layers and implantation steps.

The trench gate devices discussed in Chap. 7 rely on an overhaul of the planar CMOS design and significant effort is required. For RF circuits, trench sinkers and the reuse of shallow trench isolation in LDMOS transistors could be an attractive approach to improve performance of MMICs. Nevertheless, the benefits have to be closely investigated in order to achieve leverage over existing technology.

Integration of trench gates into an existing planar LDMOS fabrication technology appears as a promising method to reduce device resistance with limited additional effort. But again, the feasibility must be evaluated for a particular design in order to verify the suitability of the conventional topology for straight forward integration of trench gates.

When moving to high resolution manufacturing technologies, a three-dimensional channel region, e.g. FinFET technology, will be available for realization of LDMOS transistors for both power electronics and RF amplifiers at no additional cost. The FinFET technology has been introduced at the 22 nm node [2, 3]. A prediction on the availability of HV-CMOS technology using FinFETs can be derived from Fig. 10.5.

Table 10.1 Additional mask count and process modules for advanced LDMOS concepts in ICs

Technology	Masks layers	Process modules	Comment
Multi-RESURF	0/+2	0/+2	Depends on reuse of existing wells
Superjunction	0/+2	0/+2	Depends on reuse of existing wells
FinFET	0	0	Provided by CMOS technology at 22 nm
Trench gate	+2/+5	+1/+2	Trench module, HV well redesign
Integrated trench	+1	+1	Gate trench module

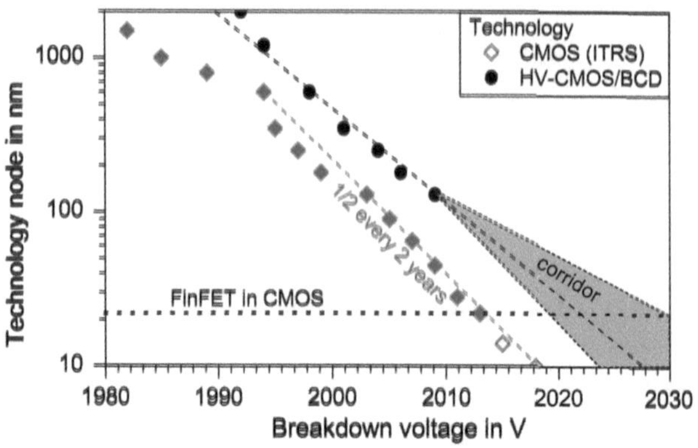

Fig. 10.5 Adoption of new technology nodes of highly integrated logic and smart-power IC processes

Data on the availability for technology nodes both in logic and smart-power ICs follow the paradigm of the International Technologies Roadmap of Semiconductors, which requires two companies to fabricate circuits at a particular technology node that year. The introduction of high voltage CMOS technology (LDMOS transistors) is typically lagging 8–10 years behind the pure CMOS technology with respect to technology node. This coincides with the average lifetime of a CMOS fabrication site before moving on to a new fab with higher resolution. The "outdated" fab can then be expanded to HV-CMOS and used to fabricate smart-power IC.

With the availability of the 22 nm technology node and the advent of FinFETs in 2011, availability of 22 nm HV-CMOS technology should not be expected before 2021. With the huge investments into the high resolution technologies and the somewhat unclear future of further More-Moore scaling, the HV-CMOS devices using FinFET technology are more likely to become a reality in 2030. At that point, LDMOS transistors with breakdown voltages below 50 V will experience a significant reduction of drift region resistance due to the three-dimensional gate shape with a larger effective transistor width. Note that FinFET technology for smart-power ICs could be introduced earlier than for the 22 nm technology node as there is experience with this device type in CMOS technology now. But this requires additional effort compared to just using the underlying planar CMOS as in present smart-power IC technologies. It does constitute to an alternative for HV-CMOS technology though because there could be a high-end technology available without the overhead of bipolar transistor processing modules in BCD.

The introduction of RF LDMOS transistors using FinFET technology can be expected to occur somewhat earlier. Integrated RF LDMOS transistors (with extended drain) are already available in 65 nm technology as of today, and the limited additional effort for their implementation could warrant extension to RF CMOS as soon as the load of the FinFET production sites for CMOS logic decreases.

10.3 Application-Specific Suitability

The previous findings on lateral power transistors are illustrated by using the applications and circuits introduced in Chaps. 2 and 3 in order to clarify how these systems might be augmented by the transistors under research (technology-push).

10.3.1 Switch Mode Power Supplies

The energy efficiency of a switch mode power supply can be increased by a reduction of on-state resistance. The achievable amount of reduction in power losses depends on the overall system, but electrical power losses in semiconductor switches is a major contributor. From Fig. 10.1 it is apparent that only GaN HEMTs are currently capable of breaking the RESURF limit. Their lack of integration with logic and the disputable reliability are limiting the rollout of this technology. The other technologies do not present a distinct advantage over the state-of-the-art topology. Certain foundry processes may benefit from the incorporation from trench-gate or superjunction technology, but there are no indications of this becoming a mainstream technology. A benefit could come from a reduction of the size of the LDMOS transistors due to a lower on-state resistance. This would increase the power density of the power switch and decrease the area required for a given transistor current. This approach capitalizes on a reduction in device area outweighing the additional manufacturing costs (see Fig. 4.14). In that case the implementation of additional process effort may be worth pursuing. However, these kinds of calculations are hampered by the difficulty in obtaining these numbers and the risks associated with the adoption of a new technology. A reduction of on-state resistance by 50 % may not be enough to warrant incorporation of a novel device technology.

A size reduction of switch mode power supplies can be achieved by incurring less power losses which reduce the size of the heat sink and higher switching frequencies which reduce the size of the inductors and capacitors. However, both approaches oppose each other because higher dynamic losses occur at higher operating frequencies.

The product of on-state resistance and gate charge is an adequate figure-of-merit to estimate the power losses incurring during a switching event. This figure-of-merit is depicted in Fig. 10.6 for different device topologies.

For highly integrated switch-mode power supplies, the RESURF LDMOS transistors currently in fabrication represent the state-of-the-art for breakdown voltages below 100 V. Trench gate and superjunction transistors do not exhibit significantly lower switching losses to warrant the additional fabrication effort. GaN HEMTs would allow for a decent reduction of switching losses and an increase in operating frequency, but they cannot be monolithically integrated with logic circuits at this point in time. For lateral SiC devices, the number of reported devices is too low to judge the suitability towards higher switching frequencies from a technological point

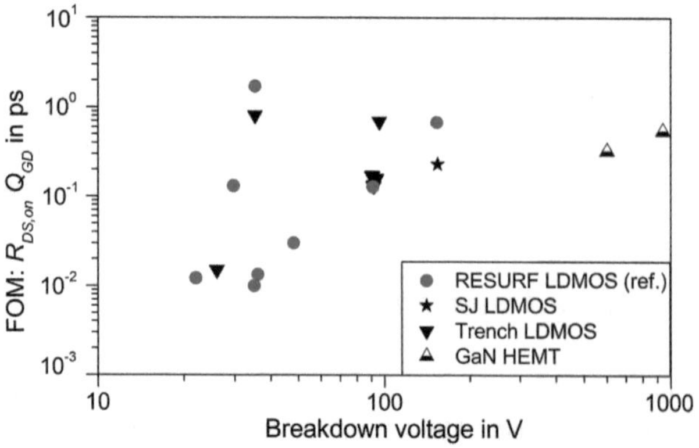

Fig. 10.6 Comparison of $R_{DS,on}$-Q_{GD} FOM for different lateral power transistor technologies

of view. Theoretically, SiC transistors should allow for lower switching losses or the increase in operating frequency.

It should be noted that the application of a source field plate similar to RF LDMOS transistors could be implemented to reduce the Miller capacitance and allow for faster switching performance of LDMOS transistors for power electronic applications. Here, the reliability of the gate dielectric and the injection of hot electrons have to be carefully considered. An improvement of electrical device properties at the cost of reliability cannot be accepted for a given application.

Based on the present data, the conclusion can be drawn that the future to planar RESURF LDMOS transistors in smart-power ICs for power electronic applications will be planar and FinFET-based (multi-)RESURF LDMOS transistors. In high voltage applications where power density is too high for monolithic integration or it is not desired for other reasons, lateral GaN HEMTs and vertical SiC DMOS transistors are competing against vertical superjunction MOSFETs and IGBTs in silicon technology.

10.3.2 Radio Frequency Amplifiers

RF amplifiers in base station equipment require high output power. Typical operating frequencies for modern mobile communication standards are in the range of 2–3 GHz. From Fig. 10.7 it is apparent that GaN- and SiC-based RF transistors allow for power gains beyond 15 dB at high power densities. Integrated silicon based devices accomplish decent power gain up to 1 W.

The desire to increase operating frequencies is linked to the implementation of higher data rates and an increase in available communication channels. The increased power gain also allows operating the RF amplifier at higher back-off, still providing sufficient power gain under the extreme peak-to-average ratios experienced in LTE, wireless LAN and other orthogonal frequency division multiplexing

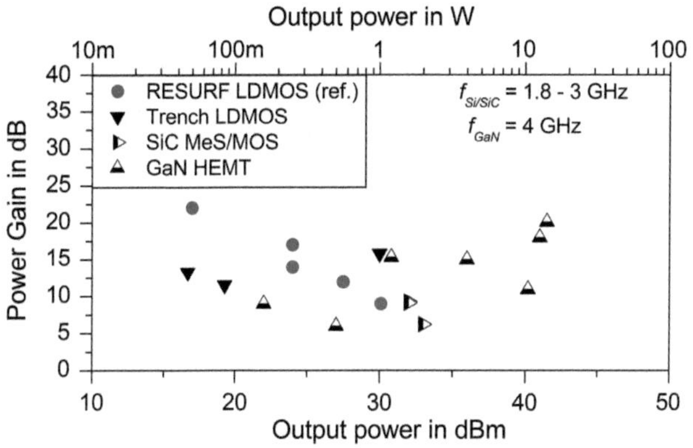

Fig. 10.7 Output power (dBm) and linear gain (dB) for power amplifiers operated at 1.8–3 GHz (data for GaN HEMTs at 4 GHz)

based applications. Similarly, a high linearity inherent to LDMOS transistors provides a high power gain because the 1 dB compression occurs at higher input powers.

The energy efficiency for telecommunication and wireless LAN in mobile systems has also to be considered. Here, a high power added efficiency enables longer uptime before recharging of the battery is required. Or conversely, higher power added efficiency means that smaller, lighter battery packs can be used.

A conclusive investigation of reliability, especially regarding gate oxide integrity and hot carrier injections cannot be given based on the published data for the different device technologies. It is clear, however, that higher electric fields in the semiconductor require additional effort in order to prevent these two effects from limiting reliability.

10.4 Evolution of Smart-Power ICs and RF Amplifiers

Based on the considerations regarding new device technologies, a conclusion towards their future in smart-power ICs and RF amplifiers can be drawn.

Presently, all major foundry processes employing LDMOS transistors rely on the RESURF principle, indicating the huge impact of this technology. With the current progress on multi-RESURF based power transistors, the superjunction and trench gate technologies are not likely to present disruptive technologies which will replace RESURF-based LDMOS transistors in general in the near future. Trench-gate type LDMOS devices will likely be implemented when the FinFET arrives in HV-CMOS technology using a FinFET-based CMOS technology. With the high effort required for silicon carbide and gallium nitride transistors logic, monolithic integration of smart-power ICs on these substrate materials is unlikely as a main stream technology as well. That being said, all of these technologies may have niche markets where their introduction warrants a significant benefit for a particular application, e.g. high temperature logic on SiC. Moreover, the demand for high efficient gate drivers for power modules employing vertical SiC power MOSFETs and lateral GaN HEMTs will readily add to the prosperity of silicon smart-power ICs. This is a More-than-Moore integration technology enabling integrated power modules as well as converters with even higher output power. In power electronics, the More-Moore road is abandoned for its lack of improvement towards LDMOS transistors. Instead, system-in-a-package solutions like wafer stacking using flip-chip or through silicon via technology present promising alternatives. This may ultimately result in highly integrated CMOS logic being paired with cheap, low resolution power devices on two separated chips being combined in a low impedance chip stack.

Regarding RF amplifiers, LDMOS transistors tailored towards high operating frequencies using a source field plate are likely to be adopted. The integration of these RF switches along with the baseband circuits into a fully-integrated RF transceiver may soon replace the GaAs-based RF output stage in mobile phones. This progress is augmented by the steady reduction in feature size by the extension of CMOS technology with RF capability following the More-Moore approach. In integrated RF applications, reduction of gate length is still a viable solution.

A summary of state-of-the-art device technologies for smart-power ICs and RF amplifiers was given in Fig. 5.19. Based on the previously introduced device technologies and semiconductor materials, Fig. 10.8 describes the prospects for smart-power ICs and RF amplifiers towards operating frequencies and voltages.

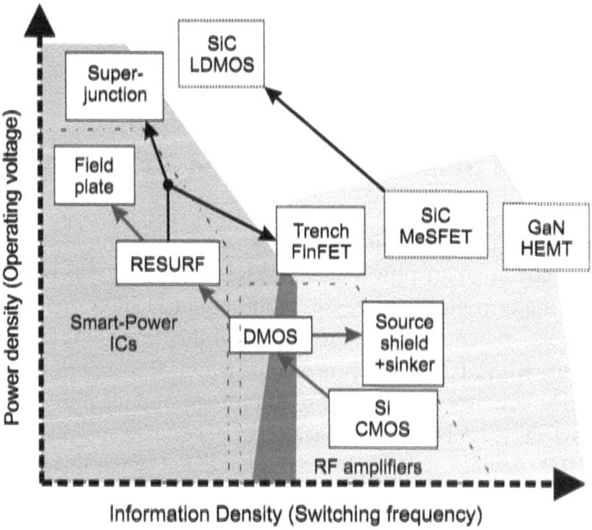

Fig. 10.8 Evolution of smart-power ICs and RF amplifiers based on novel device topologies and semiconductor materials

Still, all presented technologies open up new opportunities for particularly demanding applications. Clearly, a dense superjunction LDMOS transistors could easily break the RESURF drift region limit, and that would enable the fabrication of smart-power ICs with operating voltages in excess of 1,000 V with power densities several times higher as used today. The low transfer capacitance demonstrated using trench gate LDMOS transistors opens the door for special fully integrated, high power, high frequency class D amplifiers which may be used in energy-efficient, audio equipment in automotive and home audio applications without the need for resorting to GaN. Meanwhile, GaN HEMTs are a viable technology for integration using GaN-on-Si technology and concepts are being brought up for the eventual release of production grade epitaxial material. The steadily evolving semiconductor technologies and the tremendous leaps over this last decade on wide bandgap devices make a reliable prediction regarding their incorporation towards integrated circuits difficult.

References

1. I. Cortes, P. Fernandez-Martinez, D. Flores, S. Hidalgo, J. Rebollo, Analysis of low-voltage super-junction LDMOS structures on thin-SOI substrates. Semicond. Sci. Technol. **23**, 015009 (2008)
2. C. Auth, 22-nm fully-depleted tri-gate CMOS transistors, in *Custom Integrated Circuits Conference (CICC)*, 2012
3. C.H. Jan, U. Bhattacharya, R. Brain, S.J. Choi, G. Curello, G. Gupta, P. Bai, A 22 nm SoC platform technology featuring 3-D tri-gate and high-k/metal gate, optimized for ultra low power, high performance and high density SoC applications, in *Electron Devices Meeting*, San Francisco, USA, 2012

Index

T. Erlbacher, *Lateral Power Transistors in Integrated Circuits*,
Power Systems, DOI 10.1007/978-3-319-00500-3

MIX
Papier aus verantwortungsvollen Quellen
Paper from responsible sources
FSC® C105338
www.fsc.org

If you have any concerns about our products,
you can contact us on
ProductSafety@springernature.com

In case Publisher is established outside the EU,
the EU authorized representative is:
Springer Nature Customer Service Center GmbH
Europaplatz 3, 69115 Heidelberg, Germany

Printed by Libri Plureos GmbH
in Hamburg, Germany